Kenneth N. Walker

Airpower's Untempered Crusader

MARTHA BYRD

Air University Press
Maxwell Air Force Base, Alabama

March 1997

Library of Congress Cataloging-in-Publication Data

Byrd, Martha, 1930–1993
Kenneth N. Walker: airpower's untempered crusader / by Martha Byrd.
p. cm.
Includes bibliographical references.
1. Walker, Kenneth N., 1898–1943. 2. Generals—United States—Biography. 3. United States. Army Air Forces. Air Force, 5th—Officers—Biography. 4. World War, 1939–1945—Campaigns—Pacific Ocean. I. Title.
U53.W255B95 1997
358.4'0092—dc21
[B] 97-79
CIP

ISBN 1-58566-020-5

First Printing March 1997
Second Printing March 2005

Disclaimer

Opinions, conclusions, and recommendations expressed or implied within are solely those of the author and do not necessarily represent the views of Air University, the United States Air Force, the Department of Defense, or any other US government agency. Cleared for public release: distribution unlimited.

Air University Press
131 West Shumacher Avenue
Maxwell AFB AL 36112-6615
http://aupress.maxwell.af.mil

Contents

Chapter		Page
	DISCLAIMER	*ii*
	EDITOR'S NOTE	*vii*
	ABOUT THE AUTHOR	*ix*
	PREFACE	*xi*
	INTRODUCTION	*xv*
	PROLOGUE	*xxvii*
	Notes	*xxvii*
1	THE FORMATIVE YEARS	1
	Notes	19
2	THE SPOKESMAN FOR BOMBARDMENT	21
	Notes	38
3	MORE SCHOOLING AND COMMAND	43
	Notes	60
4	WASHINGTON AND AWPD-1	63
	Notes	83
5	THE SOUTHWEST PACIFIC AND FIFTH BOMBER COMMAND	87
	Notes	104
6	WALKER'S LAST MISSION	107
	Notes	119
7	THE LINGERING DOUBTS	121
	Notes	130

Appendix *Page*

A Walker Memorabilia 133

B Walker Articles 159

BIBLIOGRAPHY 191

INDEX 201

Photographs

Kenneth N. Walker with Paternal Grandmother 2

Kenneth's Mother, Father (right), and Maternal Uncle 3

Kenneth Works Wheat Harvest in Canada 4

Kenneth Joins the Army 5

Kenneth Walker, Flight Instructor 6

Walker and "Jenny" 7

Long's Peak 8

Marguerite Potter, Future Wife of Kenneth Walker 9

Kenneth Walker and Marguerite Potter, Wedding Party 10

Mr. and Mrs. Kenneth Newton Walker 11

Philippine Living Room 12

Marguerite in Philippines 13

Mindanao Waterfall 16

Mindanao Tree House 17

	Page
Walker in Rickshaw	18
Guarding the Trains	19
Walker (right) and Lt Col C. L. Tinker in 1931	51
B-18s over Mauna Kea and Mauna Loa in 1939	54
The "Gold Bug"	56
Generals Kenney and Walker	92
Unidentified; General MacArthur; General Blamey; General Kenney; Unidentified; General Walker	99
General Walker; Navy Secretary Knox; Unidentified . . .	99
Walker's Last Photograph	118
Missing in Action	118
Medal of Honor Citation	122
Ken Walker Jr. Receives His Father's Medal of Honor	123
Doug Walker Receives His Father's Legion of Merit . . .	124
College of Aerospace Doctrine, Research, and Education	130

Editor's Note

We are indebted to Martha Byrd for conducting the difficult and exhaustive research necessary to pull together the numerous parts that comprised the life and legend of Kenneth Newton Walker. That Walker had significant influence in the early days of airpower's rise to prominence cannot be questioned. Ms. Byrd has brought us the man behind the influence. We also thank Martha Byrd's husband, Jerry A. Roberts, for providing the information contained in "About the Author."

Since Ms. Byrd died before she could write an introduction to this biography, we asked David R. Mets, Lt Col, USAF (retired), to write one on her behalf. We are deeply indebted to him for the excellent introduction you see here.

We must also extend thanks to Douglas Walker, Kenneth and Marguerite Walker's second son. He was professional, courteous, cooperative, and helpful at all times. He furnished the photographs used in this book, and any further use of these photographs requires his knowledge and approval.

Preston Bryant

PRESTON BRYANT
Editor

About the Author

Martha Byrd was born near Morganton, North Carolina, in 1930. She was educated in the Burke County and Morganton City school systems, and graduated from the University of North Carolina at Chapel Hill in 1952. She later obtained a Master's Degree from the University of Tennessee at Knoxville. Her written works include *A World in Flames: A History of World War II* (published by Athenaum in 1970 and reprinted by Smithmark in 1992), *Saratoga: Turning Point of the American Revolution* (published by Auerbach in 1973), *Chennault: Giving Wings to the Tiger* (published by University of Alabama Press in 1987), and numerous magazine and newspaper articles. Her one novel, *A Shoebox of Violets*, was published posthumously by Laney-Smith in 1995. Martha Byrd died of cancer in 1993.

Preface

The same traits of character that marked Kenneth N. Walker's life led to his premature death. His most dominant characteristic, an inner drive that kept him at a fever pitch of intensity, was formed during a hard childhood. He joined the US Army in 1917 at age 19. Until 1928, his career was sound but unexceptional. He found his professional stride as a student at the Air Corps Tactical School in 1928–29, when he embraced the concept of the invincible bomber and made it his crusade. He served as bombardment instructor at the school from 1929 to 1934. Walker's years at the Air Corps Tactical School were critical years in the development of US air doctrine. In that process—an intellectual process that had to be primarily theoretical because experience was so limited—Walker advocated bombardment as the means through which airpower in the future would be expressed. He did extensive work on bomber tactics and plane development, the goal of which was to make the bomber capable of defending itself as well as carrying out its offensive mission. Since he envisioned defense against an enemy's air force being accomplished by bombers that would destroy the enemy's planes and facilities on the ground, he saw little value for the pursuit (fighter) arm.

Although the phrase was first voiced by others, Walker became identified with the credo, "The well-organized, well-planned, and well-flown air force attack will constitute an offensive that cannot be stopped." He believed it so fervently and advocated it so vehemently that his very conviction seemed to overcome the nagging doubts of others. Even his supporters agreed that he was "rabid" in his single-mindedness. Claire L. Chennault, his most vocal and visible opponent in the long debate on air doctrine, called him a radical with a blind spot. With limited technology, low appropriations, and an isolationist foreign policy also affecting decisions, the Air Corps moved inexorably toward a doctrine of strategic bombardment as the primary mission of an air force.

Walker's intensity carried with it a marked arrogance and self-centeredness. Between 1934 and 1941, he advanced in his career while his personal life fell apart. His first marriage ended in 1934; a second marriage lasted only a very short time. His close relationship with a third woman interfered with his efforts to become closer to his growing sons. When he left for the Southwest Pacific in 1942, he told a close friend that he had made a mess of things and might not be back.

Before leaving the States for combat duty, Walker made a further significant contribution to US airpower doctrine. Assigned to the Army's Plans Division in Washington early in 1941, he became the first staff planner in the newly established Air War Plans Division. During the late summer and fall of 1941, he and several close associates—primarily Harold L. George, Haywood S. Hansell, and Laurence S. Kuter—put together the document known as AWPD-1. Ostensibly an outline of the planes and bases and men the United States needed for the war ahead, the document was significant in that it defined a formal role for US air forces. The thrust of that role was strategic bombardment, using high-altitude daylight precision attacks to destroy key segments of the enemy's economy and capacity to wage war. The air forces would prepare the way for ground forces to invade, but the possibility was raised that the bombardment campaign might render the ground campaign unnecessary.

Brigadier General Walker arrived in Australia in August 1942 to lead the Fifth Bomber Command of Gen George C. Kenney's new Fifth Air Force. At this time, the Japanese were aggressively advancing while the Allies struggled to get on their feet and fight back. For the Fifth Air Force, planes and men were in short supply, base conditions primitive, morale low, the battle terrain formidable and terrifying as well as largely unknown. Kenney, Walker, and Gen Enis Whitehead set out to turn that around. The process required innovation and daring. Walker and Kenney soon clashed over tactics and procedures, for Walker viewed the circumstances as an opportunity to test the concepts he had spent his entire career in developing. Kenney's background was strong in attack aviation as well as engineering; he had his own concepts to test. Walker supported some of them but not others. He defied Kenney's orders on several occasions, but Kenney let the insubordination pass because of Walker's value to the command.

Innovative tactics—plus guts and sacrifice—enabled the Allies to overcome the military crisis in the Southwest Pacific in late 1942 and early 1943. The Japanese advance through New Guinea was stopped, Allied forces seized the initiative, and the Guadalcanal campaign in the neighboring South Pacific began to turn in the Allies' favor. At the end of 1942, Fifth Bomber Command was asked to blitz the main Japanese base at Rabaul, New Britain, where an unusually large concentration of enemy shipping had assembled. For what was planned as the largest bombing raid to date in the theater, Kenney ordered a night takeoff for a dawn strike.

Walker requested a morning takeoff for a noon strike. Kenney overruled him. Walker ignored Kenney's orders. He also flew as an observer in one of the lead B-17s—a direct violation of Kenney's orders that he stay out of combat. Walker's plane was one of two lost in the raid. Neither his plane nor his body was ever found.

In the sense that he died at age 45, Walker's was a career cut short. It is possible, however, that his place in history is stronger because he did not live to take part in the next phase of airpower's doctrinal evolution. His contribution to doctrine was significant but flawed: the World War II experience exploded his assertion that the bomber would not need support from fighter aircraft. We can only speculate how Walker's thinking might have changed during the course of the war. Stubbornness and pride might well have kept him from further growth. His strength, however, and the factor that earned him a respected place as one of airpower's pioneers, lay in the surety of his conviction at a time when airpower was unproved and disorganized. An untempered crusader, General Kenneth N. Walker helped ensure that the United States entered World War II with a solid foundation for the effective application of airpower. With a strong bomber and sound bombardment tactics as its base, that foundation withstood the initial trials of combat and proved flexible enough to change with experience.

Introduction

The life of Kenneth Walker spanned most of the first half of this century. It was a time of enormous change everywhere—in some ways even greater than the changes we have seen in the last half. Before we move on to Martha Byrd's examination of his life, we will here attempt to set Walker, the Air Corps Tactical School, the Army's Air Corps, and the Army itself in the context of those times.

America and the World, 1898–1943

Walker was born at the very moment of one of the greatest turning points in American and world history. We recall that was the year of the Spanish-American War, which marked America's transition from an agrarian, regional, Third World country to one of the world's greatest powers. She simultaneously defeated the once-proud Spanish navy, passed most of the other great powers in the significant indices of industrial power, and maintained her status as the giant of the agricultural world. She was rapidly changing from a rural to an urban society, and her people were becoming ever more literate. Finally, she was receiving another huge wave of European immigrants, arguably the most daring and competent that the Old World had to offer.

But sweeping changes were occurring overseas as well. In Tsarist Russia, there were faint stirrings toward industrialization and even toward democracy. Farther west, *Pax Britannica* was entering its twilight. Industrialization was proceeding much more rapidly in Germany and Japan than in Russia. Both were beginning to build great navies that would one day threaten the hegemony of Britain at sea. Economic power and military power were beginning to migrate away from London—a process that would not be matured until shortly after Walker's death, when only two great powers remained, situated out on the periphery of the Old World: Washington and Moscow. In the process, two world wars left the core of Western Civilization in economic and physical ruin. But that is only a part of the context of Walker's life. He was never stationed in Europe, and he spent the greater part of his life far to the west of the European core—far to the west of New York and Washington, in fact.

When Walker was born, the second great wave of Western imperialism had reached its peak. The European powers were completing the partition of huge portions of Africa and were attempting to do the same for China. They had previously acquired command of huge territories in the Asian subcontinent, Southeast Asia, the Middle East, and the West Indies. That very year, the first great anticolonialist power to achieve independence (the United States) switched sides and entered the struggle in the company of the imperialists—albeit with considerable hesitation, to be sure. Almost by accident, America acquired Puerto Rico plus a sort of protectorate over Cuba. But even more worrisome for many anti-imperialists, America's flag appeared for the first time in the Far East—in the Philippines. America was engaged in Asian politics as never before, and it was there that Walker's life was to come to a premature end.

Cultural Change

Probably Kenneth Walker's upbringing was more akin to that of the old America than the new. At the midpoint of his life (1920), the United States was said to have become more urban than rural. Too, the literacy rate continued its climb to near universality before he died. A huge portion of the draftees in Walker's war had not graduated from high school, but much progress had been made in public education since the First World War. Walker was not a college graduate, in common with many officers and most of the populace. The great expansion of our universities was yet to come, but there had been substantial changes in the substance of higher education during the first half of the twentieth century. The utilitarian part of our higher education system had begun its growth during the Civil War, and it continued to march with the expansion of industry and the mechanization of American agriculture. The steam press yielded cheap newspapers, and the beginnings of both electronic media and motion pictures were reducing the cultural isolation of many of Walker's countrymen—changes that were not yet found in most of the world. Also, the combination of urbanization, the telephone, the automobile, the weakening of the churches, the beginning of the liberation of women, and other things, is said to have weakened the bonds of the traditional family. One result, it is further asserted, was that divorce became increasingly acceptable in American society—especially outside of the Army (though Billy Mitchell, Douglas MacArthur, and Benjamin Foulois all had divorces that seem not to have greatly inhibited their military prospects).

Perhaps most important was a profound change in the psychological outlook of a large part of the American elite. Up to the midpoint of Walker's life, America had rarely encountered insoluble problems. She generally had a progressive and pragmatic frame of mind. Men of energy, intelligence, and goodwill could conquer any problem that came up. Such men could leave the Old World behind and build a New World where poverty and war would be a dim memory of ancient times. Diseases could be conquered with science and money. Poverty could be conquered with ingenuity, energy, and productivity. Wars could be eliminated through brotherhood and a common American heritage. But then came the horrors of World War I and its bloody stalemate in the trenches. It seemed that nothing worked anymore. The agony went on for years, and Americans were so repulsed that many of them thought war was the worst of all outcomes—anything that had the hope of evading that horror altogether, or bringing it to an end more quickly, would gain a receptive audience.

But the progressive American outlook received another profound shock in 1929. The great engine of our peace and prosperity broke down. And the great progressive, Herbert Hoover, an engineer and renowned humanitarian, seemed powerless to do anything about it. There was hunger in the land, and the old goodwill seemed spread too thin. Sigmund Freud had told us that the rational part of the mind does not drive us anyhow—that the dark corners of our subconscious were the real sources of our actions. The rules of logic could only work to give our decisions a veneer of Enlightenment rationality. The novels of John Steinbeck and other authors of the day gave further voice to this despondency—all this while the New Deal was using those who could find no work to build officers' housing on Maxwell Field, home of Kenneth Newton Walker and the Army's Air Corps Tactical School. Those were the formative years of Walker's professional career.

Technological Change

America was electrified and motorized during Kenneth Walker's short life. The world of coal and steam was much diminished; the world of oil and electrical power was booming. When Walker went to war, the entire Navy was oil-powered. Coal trucks still rumbled on the streets of New York, as the great metropolis was still largely heated by coal, but it was universally lighted by electricity—and modern lighting had spread through great parts of the rural South and West. Instant communication had been available to high commanders through the telegraph in the Civil War and through primitive radio in World War I. During the last decade of Walker's life, instant communication was spreading to more common folk through the telephone.

The transportation revolution had been under way for nearly a century when Walker was born, but it was still largely limited by terrain and the huge capital investments required for steamships and railroads. During his lifetime, the internal combustion engine released travelers from the bonds of rail and waterway. Large numbers of upper and middle class families (and some Army units) were motorized before World War II. This, too, reduced the cultural isolation of many Americans and changed personal mobility in ways that affected even the very structure of family life.

More importantly for Walker and his war, the internal combustion engine also made powered flight a practical proposition. He was only five years old when the Wright Brothers first flew at Kitty Hawk; when he died, experimental jets had already flown in Europe. This move into the third dimension was probably the most revolutionary technical change since the coming of gunpowder hundreds of years before. It was to have a profound effect on Walker's life—and on war in general—because for the first time it became practical to contemplate striking directly at an enemy's vitals without first having to conquer his fielded defense forces.

Economic Change

There were huge changes in the world's economy during Walker's time. The multiple causes need not detain us long. Suffice to say, technology, natural resources, and war pushed it all along. London had been the financial center of the universe when Walker was born, and British industry was still strong though being overtaken by that of the United States, Germany, and Japan. By the time he died, the effect of World War I had caused New York to become the world's financial capital. The Great Depression had come to Britain earlier, and its effects were more long-lasting there than elsewhere. As for the German and Japanese economies, the costs of war and bombing were temporarily rendering them helpless.

Welfare capitalism, social democracy, and communism had made enormous gains over laissez-faire capitalism in Walker's time. In his America, big labor had reacted to the depredations of big business by creating huge unions. And the government had reacted by beginning to build an economic floor which was supposed to limit the fall of those at the lower end of the economic spectrum when times were bad. Notwithstanding taxes, unions, and other constraints, American industry grew to enormous proportions and continued the development of its mass production expertise. The great automobile, aviation, petroleum, and electronic industries were near the core of these developments. When Walker flew off to the Southwest Pacific, the United States was still largely independent of foreign sources of raw materials and food—a situation much envied by Germany and most other great powers.

The Army and Its Airmen

Though he was too junior to be much involved, during Walker's early years in the Air Service and Air Corps he was witness to some grand bureaucratic battles characteristic of periods of military drawdown. It was not so much that the old horse soldiers depreciated the value of aviation, but rather that they had their own visions of the future. The real battle was over how this newly important aviation was to be used, not whether it would be a major factor. In both Army and Navy, there was a powerful strand of thought that aviation could greatly enhance the power of the older forms of battle and that it had not yet proven an ability to have decisive effects independent of the ground and naval forces.

Many Army and Navy airmen disagreed, however. In the case of the Army, the fliers asserted that soon airpower would be able to bring about decisive political results without the necessity of first conquering the enemy's army—and to do so quickly, thus reducing the total suffering. In the Navy, the airmen more gradually came to argue that the aircraft carrier would be the capital ship of the future with all the other units of the Navy existing to enhance its striking power. In their world, the argument was not settled until the midst of World War II.

In Walker's world, the more vicious argument led to the court martial of Billy Mitchell. The outcome gave some lip service, at least, to the idea that greater development of military aviation was required. However, the related Morrow Board and, 10 years later, the Baker Board both concluded that there was no clear and present danger of any serious air attack against the continental United States—and both were right. From that it was concluded, with the dissent of Jimmy Doolittle in the latter case, that there was no call for an independent air force designed to achieve independent results in war. When Walker went to war, the US Army Air Forces had gained a good bit of autonomy within the Army, but it was still a part of the Army. Almost all of the strategic bombing advocates remained cautious enough not to claim that victory could unquestionably be achieved without the help of the Army and Navy.

It is worth noting that though Walker and almost all of his Air Corps cohorts forever complained that Army airpower was being starved, that was not entirely true. On the eve of World War II, it was shown that from 1926 to 1940 the research and development money granted to all the other combat branches of the Army amounted to only about 60 percent of that devoted to the Air Corps by itself. The Army had long ago folded its infant armor organization back into the infantry and its antitank guns were absolutely primitive compared to those in Europe. In the National Defense Act of 1920, the Congress had cut Army manning to half of what had been recommended—and then only funded it to provide about half of that. After all, the "Great War" had been a war to end all wars. The point is, though, that if the Air Corps was undernourished, the rest of the Army was really starved. For all of that, though, the interwar Army was double the size of the one that had existed in 1914.

Army Air and the Navy

Mitchell was convicted, and he certainly was guilty as charged. Afterwards, the bureaucratic conflict between the airmen and the Army General Staff was held within more circumspect bounds, if only because of the Mitchell example and the fact that the General Staff could greatly influence the funds made available to aviation.

However, one of Mitchell's main themes had been that navies were increasingly obsolete and that airpower could defend both coasts more effectively and at much lower costs than could navies. For the airmen, the Navy remained the principal bureaucratic enemy right up to Pearl Harbor—and afterwards they never tired of pointing out that Mitchell had accurately predicted the Japanese attack (and the Navy itself had practiced similar mock attacks on Pearl Harbor in the early thirties). Further, these feelings did not disappear with the coming of war, particularly among the forces deployed to the theater in which Walker died—the Pacific. For most of the prewar period, though, these bureaucratic struggles were fought out at levels far above that of Kenneth Walker. Mitchell himself from 1919 forward was a great boon to Adm William Moffett, head of naval aviation. Moffett was able to use the airman as a bogey to coerce the battleship admirals into releasing more money to naval aviation than they might have. His argument was that if the Navy did not pursue aviation development with vigor, then Mitchell would take it away from the sea service. By 1941, US naval aviation led the world in most of its functions and technologies. But though Walker himself was no more than an observer of those external battles, he was directly involved in a struggle within the air arm at the Air Corps Tactical School.

Bombers Versus Fighters

Much has been made of the internal struggle at Maxwell Field in the early 1930s between fighter pilot Claire Chennault and a very few allies against what is usually painted as the majority conventional wisdom of the bomber barons. Equally usual is the implication that Chennault was right and the bomber people, including Kenneth Walker, were wrong—especially in the notion that the bomber could get through in daylight with acceptable losses to achieve decisive damage to vital targets.

During Walker's younger years, in World War I, airplanes were at first used for reconnaissance and artillery spotting. This was one of the factors quickly leading to the stalemate in the trenches—the offensively minded general could never mass the required numerical superiority at the decisive point without his enemy finding out about it. Moreover, the spotting so enhanced the accuracy of artillery fire that the new fragmenting rounds were deadly against offensive troops necessarily out in the open. So the demand for command of the air first came from the ground generals. The airmen themselves were quick to take up air superiority as their first priority mission; it remains so still.

Mitchell came back from Europe in 1919 with the idea that air superiority was best achieved with pursuits (now called fighters) and that had to remain the priority mission—so, as I see it, the fighter units were the elite of the Air Service during Walker's formative years. Meanwhile, the Italian general Giulio Douhet was arguing that air superiority could best be achieved by bombers through attacking enemy airpower while it was on the ground—in a world without radar, one could not find the enemy attackers in the big sky soon enough to stop them. As the decade of the 1920s wore on, the idea that the air battle would be a necessary part of the struggle for command of the air (Mitchell himself had seen some virtue in attacking airpower on the ground where possible) retained many supporters. But the Douhet-like idea increasingly gained ground, though he was seldom credited with influence.

From the First World War, airmen almost universally agreed that fighter escort would be a nice thing for bombers to have. However, they most usually doubted that it would be technologically feasible to build a fighter with the tankage that would permit it to duplicate the range of the bomber and yet retain enough agility to tangle with enemy interceptors. The escort would necessarily be heavier and larger than the interceptor, which did not have to fly so far to get back to its base. Chennault himself was one of those who asserted the impracticality of the escort fighter. It is likely that the most avid bomber advocates, realizing that escort fighters were generally impractical (remember Iwo Jima),* wished so much that the bombers could defend themselves that they persuaded themselves it was feasible. Many will disagree, I suppose, but I believe they were right—in the absence of radar, and it would have taken a miracle of foresight to predict the coming of radar in a world where the German army was still horse-drawn. I offer as support the record of F-117s in 1991 against an air defense system much more formidable than anything that could have been put up by the Germans and Japanese—the F-117s did indeed always get through and the most important reason was that radar had been factored out of the equation with stealth technology.

The point is that when Walker and his cohorts were arguing against Chennault at the Air Corps Tactical School in the early 1930s, radar was not a part of the equation. I leave it to the reader to judge whether he or any of the other bomber advocates should have been expected to predict the coming of radar so soon.

Germany First

One of the most important factors conditioning Walker's war was the grand strategy of the alliance. Even before Pearl Harbor, the Americans and the British decided that Germany would have to be defeated first if war should come with all the members of the Axis. This guaranteed that the Pacific War in general would have a poor priority for manpower, equipment,

*Six thousand Marines died in taking the island, which was needed in large part for a base from which P-51s could escort B-29s to Japan.

and leadership. Moreover, though Adm Ernest King was partly responsible for that grand strategy, he did about everything he could to build up the Navy's part of the war in the Central Pacific—and as Walker went to MacArthur's part of the region in the Southwest Pacific Area, MacArthur's part had to compete with Nimitz's part for resources. Thus, MacArthur's constant complaint on that issue had some foundation in reality.

Walker's War

MacArthur's war before Kenneth Walker's arrival was almost an unbroken string of disasters. As Yamamoto had predicted, the Japanese ran wild for the first six months. After their great victory at Pearl Harbor, Nagumo and his carriers bashed Darwin, Australia, on the way to driving the British Navy out of the Indian Ocean. Meanwhile, though it took a while, MacArthur was driven out of the Philippines in humiliation. In the process, he fired his first air commander, Lewis Brereton, and replaced him with George Brett. MacArthur was no happier with Brett, who was sent on to other regions. The next replacement was George Kenney, who arrived in the summer of 1942—about the same time as Walker.

Kenney knew Walker well. They both had been on the faculty of the Air Corps Tactical School at the same time, with Kenney in charge of the "attack" (something like today's "tactical") part of the curriculum. Kenney was therefore fully cognizant of the strategic bombing theory even before he left, but later was the operations chief in the staff of Gen Frank Andrews' General Headquarters Air Force (GHQ AF). The GHQ AF has been cited as a major step in the maturation of the American theory of strategic bombing and, organizationally, en route to an independent Air Force. Though Kenney and Walker had much in common, the latter was not Kenney's leading lieutenant—that honor belonged to Enis Whitehead, who was doing the day-to-day leading of Kenney's combat operations. Walker, by then a brigadier general, was Kenney's bomber commander. He served as such during the initial six months of MacArthur's campaign in New Guinea.

With that much as introduction, we shall now turn to Martha Byrd's description of Kenneth Walker's life and times. She was eminently qualified to carry out the task. Born while Walker was assigned to Langley Field, before the Air Corps Tactical School was moved to Maxwell, she authored books on the Second World War, the Battle of Saratoga in the American Revolution, and the life of Claire Chennault. I trust the reader will agree that she was blessed with the writing style and historical background to write a fine work about Walker and his times. I trust also that the reader will share my regret that she did not survive to finish the work herself.

David R. Mets
School of Advanced Airpower Studies
Maxwell AFB, Alabama

Prologue

The road to the Medal of Honor is not followed without determination, without courage, without a measure of personal pain. Maybe Ken Walker sensed, on that day in July 1942 when he left for the Southwest Pacific, that his dedication to his career had led him to pursue the road beyond turning back. "I've made a terrible mess of things here," he told Bob Pearson, a lifelong friend who shared his doubts and dreams. "I doubt if I'll be back."[1]

The mess that weighed on Walker's mind was not professional, but personal—two divorces, three sons who scarcely knew him, other relationships broken or unfinished. In contrast, his career formed a neat package, each orderly layer stacked upon the preceding to shape a solid entity that could withstand the erosion of time and criticism.

Not that Walker's professional life was without flaw—though he was a leading thinker during the formative years of US airpower. He made errors in judgment, errors of commission and omission, like other airpower advocates. No fault or failure, however, resulted from lack of effort on his part. To the contrary, among his Air Corps colleagues his name came to be equated with intense—almost furious—pursuit of an idea or task.

"That's why he didn't come back," mused Mamie Lee Andrews, a close friend of many years. "He loved his work."[2]

Notes

1. Scott Pearson, Denver, Colo., telephone interview with author, 7 February 1991.

2. Mamie Lee Andrews, Chatham, Mass., telephone interview with author, 28 August 1990.

Chapter 1

The Formative Years

Kenneth Newton Walker was born in Cerrillos, a small New Mexico town south of Santa Fe, on 17 July 1898. His mother was Emma Overturf Walker, descended from one Samuel Overturf, who was born in Pennsylvania in 1787 and moved to Indiana in 1817. Samuel's son Newton, for whom Kenneth Newton Walker was named, married Mary Alice Wade at the end of the Civil War. Newton and Mary Alice Overturf moved to Nebraska in 1888. Emma was one of their nine children.[1] In due time, Emma married Wallace Walker. We know little about him except that he grew up an orphan, raised by a family surnamed Walker. When and why Wallace and Emma Walker moved to Cerrillos, New Mexico, remains unknown. The assumption is that Wallace went there to work in the nearby gemstone mines.

A lone surviving picture of Wallace Walker shows him nattily dressed, with shoulders back and hands in pockets, and wearing a black derby with the confidence of a man of the world. He left Emma and their son when the boy was quite young. Emma raised young Kenneth alone, and she did it by hard work at whatever job she could find. A slender, attractive woman, she was devoted to her son—and he to her. Kenneth grew up to look much like the derbied figure in the photograph, but his opinion of his father was harsh.[2]

Sometime during Kenneth's early childhood, the family moved to Denver, Colorado, where Kenneth attended the Maria Mitchell School (1905–08) and later the Columbian School (1908–12). While mastering his reading and arithmetic, Kenneth also acquired an enduring friendship with classmate Robert Pearson. Bob Pearson and Ken Walker shared boyhood games, adolescent struggles, and maturing interests. In the process, they developed one of those rare symbiotic relationships that grow stronger with the years.

In 1912, Kenneth attended Central High School in Kansas City, where he belonged to a Boy Scout troop and channeled extra energies into basketball, football, wrestling, and boxing.

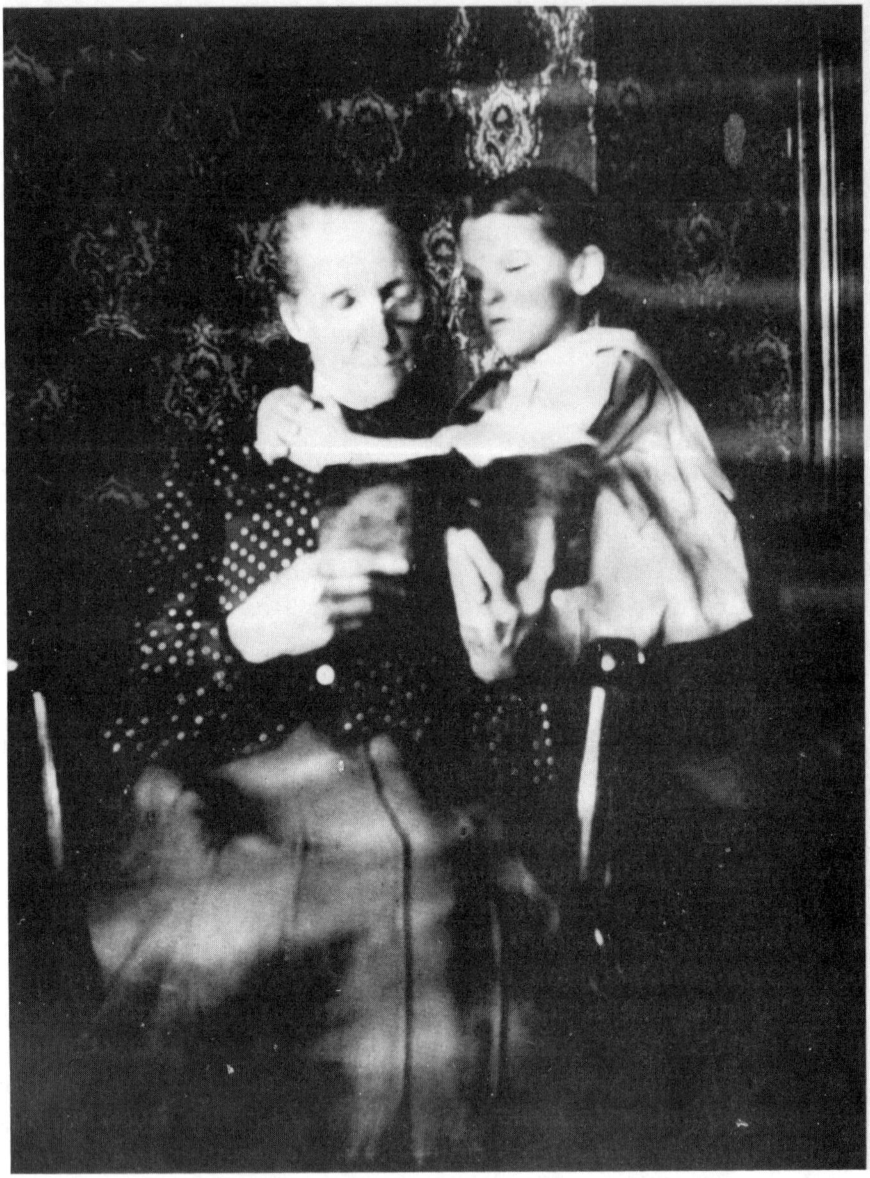

Kenneth N. Walker with Paternal Grandmother

One summer—probably 1914—he worked with the wheat harvest in Canada. From the fall of 1913 until June 1915, he was enrolled in the Omaha High School of Commerce, from

Kenneth's Mother, Father (right), and Maternal Uncle

which he graduated. Emma Walker's parents lived in Omaha; possibly Emma, by that time a single parent, had moved there to be near them. By 1917, however, Kenneth was back in Denver, where he attended the YMCA Night School between January and June. He graduated from this course also, then took an extension course in business administration from La Salle University. This sparse outline suggests that he was pursuing any open door in seeking his niche. He had been an executive trainee in a Denver corporation for a short time when he enlisted in the Army on 15 December 1917.

He might have been looking beyond World War I and welcoming the opportunity the Army offered for a career. A photograph taken in his new uniform shows an immature but confident face, feet planted firmly on the ground, and hands held behind a solid, somewhat stocky body.[3]

Kenneth Works Wheat Harvest in Canada

Walker's World War I career was unexceptional. From the Aviation Section of the Signal Enlisted Reserve Corps, he went on active duty with the Aviation Section Detachment at the University of California. June 1918 found him at California's Mather Field, learning to fly at the Air Service Flying School. At the time of the armistice in November 1918, Walker was honorably discharged from the Reserve to become a second lieutenant in the US Army Air Service, that branch of the service with the most glamour but the least prestige and money. However, its brief but dramatic role during the war had convinced a few visionaries that herein lay the warfare of the future. Walker and the Air Service's 1,170 other officers considered themselves lucky. The sky was full of challenge, and they had plenty of leftover DH-4s plus the freedom and time to fly them.[4]

Kenneth Joins the Army

Walker began his Air Service career much as did the other officers of his day. First came Flying Instructor's School at Brooks Field in late 1918, then duty as a flying instructor at Barron Field. Instruction was by Gosport, with Walker riding in the seat behind the student and barking guidance/criticism through the one-way Gosport communication tube that allowed the student no opportunity to talk back.

In March 1919, he was assigned to historic Post Field at Fort Sill, Oklahoma. Today a National Historic Landmark and headquarters of the US Army Field Artillery, Fort Sill was established by Gen Philip H. Sheridan in 1869 as a base for the southwest Indian wars. Geronimo, the last great Apache chief, lived there between 1894 and 1909; he is buried in the Old Post Cemetery. Once the Indian wars were over, Fort Sill began to serve the artillery branch and, later, the Air Service. During 1918, the School for Aerial Observers and the Air Service Flying School were built at adjacent Post Field, where Walker served for four years. His assignments included pilot, instructor, supply officer, and post adjutant. In 1922, he added combat observer to his command pilot rating.

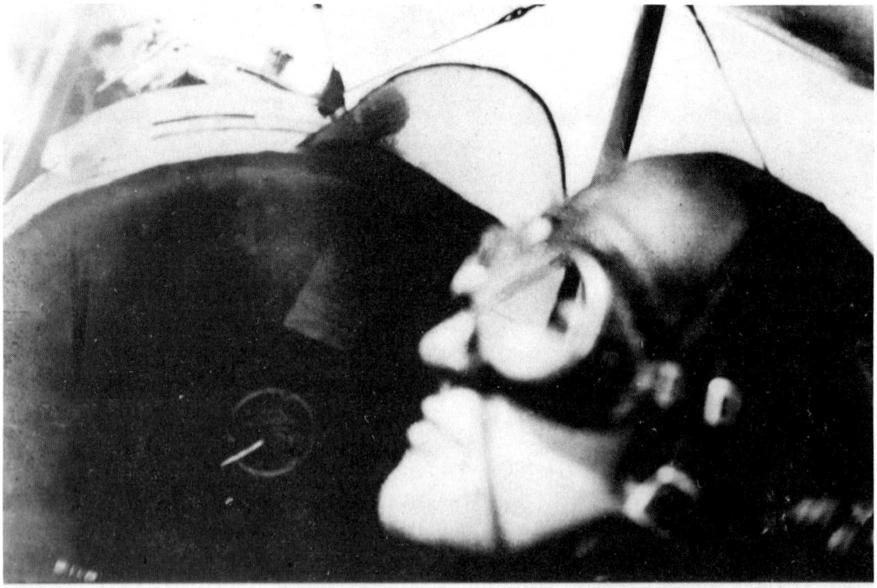

Kenneth Walker, Flight Instructor

Walker and "Jenny"

Walker was one of a select few who took airpower seriously during those years. Airplanes and flying were so new—and so remote from the experience of most people—that they inspired awe if not terror. The public was being introduced to flight through air shows, stunt flying, and competitions, all of which contributed to the image of flyers as glamorous daredevils—an image Walker was not averse to exploiting. Like most pioneer aviators, he possessed generous measures of showmanship and audacity, qualities that he revealed on Labor Day, 1919, when he treated a few people to an unofficial demonstration of just what a small biplane could do. More than likely, he had signed out for a routine cross-country flight—if, indeed, in those days of few regulations, he signed out at all. In any event, he first flew to Denver and landed at a small dirt strip east of town, about where Stapleton Airport is now located, to pick up his friend Bob Pearson. Pearson worked in the mining machinery business, but his avocation was the air. "A frustrated flyer," Pearson's son called him; he spent a good bit of time building planes and experimenting with aeronautical engineering. One of the bonds of friendship between Pearson and Walker was their mutual deep fascination with the air.

On this Labor Day, Walker flew his friend up to Long's Peak, one of Colorado's 14,000-foot mountains near Estes Park. In a DH-4, which had a service ceiling of 19,600 feet, they flew all around the top of the rugged, impressive summit, taking pictures to verify their accomplishment. They came down over Boulder and the campus of the University of Colorado at rooftop level, still taking pictures. On to Denver, where they flew up 16th Street at the level of the tops of most of the buildings—actually below the level of a 400-foot tower. Onlookers gasped and held their breath as the plane flew toward the capitol at the end of the street. Around it they flew, at dome height. By this time, others besides Pearson had cameras working overtime. Finally, having given onlookers a spectacular show, Walker flew back to the dirt strip, touched down just long enough for Pearson to jump out, then took off again with a flamboyant wave of the hand. The image of the aviator—glamorous, fearless, forging upward into new frontiers—was an image Walker embraced with ease.[5]

While stationed at Post Field, Walker met and courted Marguerite Potter, a sociology graduate of the University of

Long's Peak

Oklahoma in Norman. Beautiful and lively, a sorority member and one year the campus "Sooner Queen," Marguerite also came from a family of westward pioneers. Her mother, Margaret Hiller, could follow her lineage back to its first American members in New Jersey in 1740. Her father, Hosea Potter, was descended from Palmer Tingley, who came to America in 1635. Later generations of the Potters settled in Vernon County, Missouri. Hosea left Missouri in 1901 to be one of the first settlers in the new Oklahoma territory and the newly formed town of Indiahoma, where he opened a store and served as postmaster.

Marguerite Potter, Future Wife of Kenneth Walker

Marguerite grew up knowing what the word "frontier" meant. In later years, she described her childhood as "a safe, secure life," with freedom, a pet donkey, and no unmet wants. Their small town had board sidewalks and Indians.

Marguerite attended a one-room school through the eighth grade, then boarded in Lawton, where Geronimo had been a familiar figure, to attend high school. Upon graduation from high school, she entered the University of Oklahoma at Norman. The ample supply of young men at Fort Sill meant plenty of dates and fun for Marguerite. She and Kenneth married in September 1922 after a two-year courtship. Their honeymoon was courtesy of the US Army: on 12 December 1922, the Walkers left the United States via a troop transport ship for a two-year tour in the Philippine Islands.[6]

Kenneth Walker and Marguerite Potter, Wedding Party

From the semidesert of Fort Sill to the humidity of the Air Service base at Camp Nichols (now the site of the Philippine National Airport) was a six-week journey in time but a quantum leap in facilities and amenities, even for someone who had grown up on the frontier. Potholes dotted the coast road from Manila to the base, six miles south; the stench of garlic, fish, smoke, and waste permeated the air. Flies abounded by day, mosquitoes by night. At the base itself, the assorted buildings had the gray and rotting look of

Mr. and Mrs. Kenneth Newton Walker

impermanence; a few planes were parked around two rusting metal hangars at the edge of the small sod field. An occasional palm tree eased the eye, but swamp surrounded the base except where the enterprising natives had thrown up small villages dominated by bars, brothels, and nightclubs.[7]

Since the base offered housing for senior officers and enlisted men only, junior officers and their families lived in Manila, a developing city where surroundings were relatively pleasant. The Walkers settled into a Spanish-style house across the street from the Philippine University and next door to Harvey and Virginia Shelton. The Walkers and Sheltons became firm friends. The two Shelton boys, Wirt and Douglas, later became pilots largely "because of Uncle Ken," and the

Walkers named their second son Douglas because they liked that name and the associations it carried for them.[8]

Philippine Living Room

The Army's Philippine Department followed the pattern of other posts abroad during those years: the Americans formed a close mutual support group and created a social life closely reflective of that in the States. The Walkers took an active part in this social process, entertaining at the Polo Club or joining the other officer families to give a despedida dinner at the Manila Hotel. Recreation could, however, be considerably more basic and unsophisticated; for example, the Philippine national sport was cockfighting—and Shelton's 31st Infantry was known as the "Thirsty-First," probably for good reason. There were ball teams on the base, of course, and some of the officers joined the golf club at Fort McKinley, the main Army base. The boxing matches at Fort McKinley also drew large crowds—the Air Service had a "very promising" middleweight and a flyweight upon whom high hopes rested.

The Air Service in the Philippines was commanded by Maj George E. A. Reinburg. Both he and his wife were popular and well-liked. Duties were not strenuous. Walker's first assignment was commanding officer of the Air Intelligence

Marguerite in Philippines

Section at Camp Nichols. In April 1923, he became property officer of the Philippine Air Depot. At other times, he served as supply officer, adjutant, and depot inspector. In 1924, he was assigned to the 28th Bombardment Squadron. Overall, his tour was pleasant. There were practice formation flights, familiarization flights, and reconnaissance flights, illustrating the relatively untried status of aviation at that time. Walker and another lieutenant once carried out a radio mission by taking off when a transport ship was 15–20 miles out in the

China Sea; they flew over the transport and radioed messages to personnel on board.

Another day, Walker took part in a cross-country pigeon-training mission; the plane released the birds at Corregidor Island and at Clark Field. The pilots returned safely; presumably, the pigeons did also. When the transport *Thomas* entered port on its quarterly visit, formation flights helped celebrate the occasion. Walker usually flew a DH-4 but, after assignment to the 28th Bombardment Squadron, he sometimes flew Martin bombers. At least one of these was the NBS-l, designated a short-range night bomber. And he made at least one flight to Corregidor piloting a seaplane.[9]

Since the summer rainy season brought severe rains that could be hazardous to flight, not all the flights were simple routine. Walker and a captain took off in two planes at nine o'clock one morning on a flight to Mindoro. They encountered heavy rainstorms, became separated, and Walker returned alone the next day. Fortunately, the captain made it safely back on the following day.[10]

On 22 August 1923, on an airfield that had been turned into mud by the rains, Walker suffered the indignity of an accident that demolished the right wing, tail assembly, landing chassis, and propeller of a DH-4BPl. Taking off on a testing flight, he had a much longer run than usual because the muddy field made it hard to keep the plane headed straight. Realizing that he was probably going to hit a garage alongside the field with his right wing, he lowered the left wing in an attempt to clear the building. Walker knew what he was doing—he had 1,340 flight hours by this time—but his maneuver did not work, probably because he lacked sufficient speed. Neither he nor his passenger was hurt.[11]

In December 1923, Walker was one of six men, three of them pilots, who made an extended reconnaissance trip to Zamboanga, the heart of Moro country on the southern island of Mindanao. Because this was a part of the world and a culture relatively unknown, Walker wrote a lengthy account of their adventure for the *Air Service Newsletter*. They could not fly to Mindanao, for the distance required refueling and the airfield at Iloilo on Panay had not yet been completed. Instead, they went by steamer, their three DH-4Bs lashed to the hatch.

After a stop at Cebu, where they saw the Leper Colony and the place where Magellan lost his life, they steamed into Zamboanga early on 9 December. It is "a beautiful little city," Walker wrote, "with white stone buildings, clean streets flanked with palm trees and tropical plants." They were welcomed by the commander of Pettit Barracks, Maj Allen S. Fletcher. The Zamboanga Carnival was in progress and the scene was exotic, with throngs of Moros dressed in colorful and dramatic costumes.

A landing field—400 by 75 yards—had been improvised for them on the golf club fairway. Fletcher cut down a large tree to improve their landing approach and provided tents and guards for them. From this base, the men flew daily reconnaissance flights over the mountains and jungles north of Zamboanga. Officers from Pettit Barracks sometimes accompanied them in order to study proposed ground routes.

One Sunday they flew to Camp Kiethly, a constabulary (Philippine police force) camp in Lanao, flying over "Sibuguey Bay, a water hop of 80 miles, over trackless jungles and a number of beautiful lakes." The field on which they landed was short and swaybacked, "but as it had been cleared off in two days by Moros armed with bolos it was a work of art." They were met by constabulary officials and some 1,000 Moros, impressive-looking men who wore colorful head cloths and sarongs. They were fascinated by the airplane, for they had never before seen one. Major Fletcher, who spoke their dialect, was asked by one of them if he had flown high enough in the great bird to see heaven. Fletcher first assured him he had, then decided not to carry his joke too far when he remembered that these Moslems believed a bevy of 40 virgins awaited every true believer and that the best way to get to heaven was to kill a Christian.

On Monday, the Air Service men were taken by automobile to see Maria Christina Falls (512 feet high) and then through a deep canyon matted with tropical foliage to another constabulary camp. They returned to Lanao to learn that 31 Moros had been killed in a fight with the constabulary soldiers.

Later in the week, when they flew south to Cotobato, Walker noted, "The jungles over which we flew appeared from the air

Mindanao Waterfall

to be impassable and it can be readily seen that the average speed of a mile an hour through the underbrush and thick tropical growth would be good time indeed." They made several long hops over water, and part of the "entertainment" for the pilots was stories of sharks. Walker soberly observed that it "sometimes appeared a long way to land."

Throughout their trip, the men collected data on emergency landing fields and places that offered favorable sites for permanent fields should the need arise. They were scheduled

Mindanao Tree House

to return to Manila on Christmas day, but a typhoon was reported to the north and the trip was delayed until the 27th. The intermediate field at Iloilo being completed, they landed there en route, reaching Camp Nichols on the 28th.[12]

Marguerite apparently had no word from Walker between Christmas day, when she expected his return, and his actual arrival three days later. She took part with the other base wives preparing for a party at the Manila Army/Navy Club, trying hard to master her fears because her husband had not returned. She went to the party with the Sheltons, but she was in tears for much of the evening. Young Douglas Shelton remembered it as a sad Christmas. Walker appeared a few days later, hale and hearty, and teasingly explained his absence by saying that he had decided he wanted to see Australia and had simply flown down there to have a look. Young Douglas bought the story, but he was left with a lasting impression of Walker's apparent lack of concern for his family.[13]

In the summer of 1924, the Walkers made a trip to China, the customary highlight for those serving a Philippine tour. They enjoyed the usual tourist sights, such as the palaces and temples of Peking, but they most enjoyed being entertained by several Chinese families. When stationed at Post Field, Walker had befriended a Chinese officer, Captain Shen. Shen was now stationed nearby, and Walker got in touch with him. Shen flew to Peking and took Walker to lunch with General An, chief of the Chinese Air Service. After lunch came a shopping trip for gifts, followed by an evening at the Chinese theater. Marguerite joined them for the latter and particularly enjoyed the company of several of the Chinese officers' wives, even though they spoke no English. The following evening, they were entertained at dinner in the Ans' home. Marguerite

Walker in Rickshaw

Guarding the Trains

considered the entire experience very special and noted that she had never been treated so cordially.[14]

The Walkers' Philippine tour was pleasant and uneventful. When they returned to the States in February 1925, however, they jumped into a boiling pot. Aviation had the military world in a state of ferment, and Walker stepped forward to be one of its future movers and shakers.

Notes

1. Information provided by various members of the Overturf Family.

2. Kenneth N. Walker Jr., Berkeley, Ontario, interview with author, 8 March 1991; Douglas Walker, New York, interview with author, 19 January 1992.

3. Premilitary schooling from family records. K.N. Walker Jr. thinks Wallace Walker left Emma when their son was twelve, which would have been 1910. The family photograph album confuses Kenneth's age; it shows him in a World War I uniform and says "age 17," although he was 19 when he enlisted.

4. Walker's 201 File. The Air Service Reorganization Act of 1920 authorized up to 1,516 officers and 16,000 men out of a total for the entire Army of 280,000. Since appropriations went down each year between 1921 and 1924, the authorized strength was not attained. Edward O. Purtee, *History*

of the Army Air Service, 1907–1926 (Wright-Patterson AFB, Ohio: Historical Office, Air Materiel Command, 1948), 124–28.

5. Scott Pearson, Denver, Colo., telephone interview with author, 7 February 1991.

6. Marguerite Walker, videotaped interview. Information on Potter and Hiller families was included.

7. Description from *Air Service Newsletter* and George W. Goddard, with DeWitt S. Copp, *Overview: A Life-Long Adventure in Aerial Photography* (Garden City, N.Y.: Doubleday, 1969), 169–73.

8. Douglas C. Shelton, Little Rock, Ark., to Douglas P. Walker, letter, 18 January 1991.

9. *Air Service Newsletter* (ASNL).

10. Ibid.

11. Accident Report 200.3912-1, US Air Force Historical Research Center, Maxwell AFB, Alabama.

12. ASNL, 7 March 1924; a typescript is among family papers.

13. Douglas C. Shelton to Douglas P. Walker, letter, 18 January 1991.

14. Marguerite Walker to her parents, letter, 11 May 1924.

Chapter 2

The Spokesman for Bombardment

The midtwenties marked a turning point in aviation. During the years following World War I, the flamboyant Gen William "Billy" Mitchell kept the question of military aviation alive. He spoke out on how airpower should be employed and administered, stepping all over the Navy's toes by demonstrating that bombs could sink battleships. He overstepped the bounds of discretion in 1925, accusing the Army of gross mismanagement of the air arm, and was court-martialed. Shortly afterward, he resigned from the Air Service, leaving behind unfinished business but a two-part legacy: (1) the firm conviction that airpower had great portent for the future of military affairs and (2) a precedent of aggressive advocacy.

In 1926, the Air Corps Act officially acknowledged a potential for the air arm and gave military aviation a little room to grow. By that time, the World War I stocks of DH-4s and Liberty engines had been depleted and the road was open for engineering advances and new plane designs. Charles A. Lindbergh flew solo from New York to Paris in 1927, capturing the public imagination with this marvelous new frontier of flight. Aviation was ready to move, and Ken Walker was in the right place at the right time with the right talents and experience to drive it along. Military airmen now needed to decide what role airpower should play and how air forces might best be organized. They also needed to move the technology along and develop doctrine for how this new force could and should be applied.

In the summer of 1925, the Walkers moved to Langley Field, Virginia, which was to be their home until the summer of 1931. One of Walker's new duties was to serve on the Air Service Board, which was charged with making "recommendations looking to the improvement of the Air Service"; but the board accomplished little with its limited personnel. Walker served for a brief time as adjutant of the 59th Service Squadron, taking charge while its commander had a month's leave. Then, in October 1925, he became commanding officer of the 11th Bombardment Squadron.

This was an exciting time to be in bombardment, for significant progress was under way on all sides. Improvements in planes, bombs, and bombsights encouraged serious work on tactics, and airmen were beginning to see bombardment—rather than pursuit—as the cornerstone of airpower. When Walker and other Air Service officers attended the National Air Races in Philadelphia in 1926, the Liberty Bell trophy competition for light bombing planes was reported as the most sensational event of the first five days. Walker himself, piloting a Huff-Daland Pegasus with an 800-horsepower (hp) Packard engine, came in second in the race, completing the 120-mile course in 1 hour, 1 minute, 9.33 seconds—an average of 119 mph.[1]

A new vision of what airpower could accomplish was emerging. One of those who shared the vision was Maj Hugh J. Knerr who, in 1927, became commander of the 2d Bombardment Group at Langley Field. Knerr brought to the 2d, one of only three full air combat groups in the United States, not only his vision but also the grim intensity of the practical midwesterner. He found the 2d, with its Keystone and Martin B-2 bombers, stagnating under indifferent leadership. Pursuit was considered both more glamorous and more important than bombardment at that time, but Knerr saw many things to do and set out to do them.

When the 11th Bombardment Squadron was reassigned to March Field, California, Walker stayed at Langley as operations officer for Knerr's 2d Group. Knerr came to depend heavily on Walker's organizational skills, his sharp and brilliant mind, and his inherent planning ability.[2]

Walker could not have asked for a more valuable experience than working under Knerr, who did not hesitate to rattle the establishment. In August 1928, he led his group on a 41-hour cross-country flight to demonstrate both the airplane's capability and the Army's paucity in fuel allotments. Before long, Knerr and Walker had a reorganized and streamlined group. They practiced a number of different bomber formations to provide mutual fire support and cross fire for defense against enemy pursuit planes. Knerr also put Walker to work on bombsights—and the 2d was soon trying out a system of strings running from the bomb bay to the pilot's arms, which the bombardier could pull to guide the pilot. When some aircraft engineers who had heard about the bombsight research came to the field to talk things over, Walker briefed them. Before long, design competitions were under way.

As early as 1927, Knerr had urged the development of a bomber that could carry a load of 1,000-pound bombs and cruise at 150 mph—a feat that seemed out of reach at the time. By 1931, however, the potential of the Norden bombsight and the Martin B-10 bomber were dramatically altering ideas on what bombers could accomplish.[3]

Stubborn, zealous, an innovative thinker convinced that airpower would be a tremendous strategic force, Knerr helped to broaden Walker's thinking. One of Knerr's convictions was that supply and maintenance would prove critical in future combat and that the primary power of the air arm lay with bombardment's ability to take away an enemy's beans and bullets. The reverse of the coin was that aerial transport must supply its own forces. One of his achievements was to cobble together the 1st Transportation Group, forerunner of World War II's Air Transport Command. Walker was impressed.[4]

When Walker was assigned as a student to the Air Corps Tactical School at Langley Field in December 1928, he was ready to take full advantage of the experience. The school had been founded in 1922 with a largely tactical, technical, and administrative curriculum. By 1928, tactics had become proportionately a larger emphasis. Fundamental doctrine had become a growing issue, however, and a subdued excitement permeated the school. World War I seemed to teach that pursuit dominated the air arm because pursuit had attained and held the air superiority that enabled the ground forces to prevail. This concept gave pursuit the most important role in an air force, but it also made aviation nothing more than an adjunct of the surface forces.

Mitchell, however, was convinced that strategic bombardment was the proper role of the air force—even though there had been limited opportunity to test it in practice. During the twenties, he led Air Service thinking in that direction. Under this concept, aviation was not merely a new weapon; it was a revolutionary new power that could alter the way wars were fought by striking directly at an enemy's will to resist. By the late twenties, the leading edge of Air Corps thought was that the objective of war was to overcome the enemy's will to resist and that defeat of an enemy's armies and navies was but one means to that end. Air force, however, provided a way of attaining the true objective without the long, bloody battles necessary to defeat armies and navies. The proper mission of the Air Corps, therefore, was as a separate and strategic force, not as a tactical weapon subordinate to the Army.[5]

This view was shared by few outside the narrow circles of aviation, however. Neither the organization nor the equipment nor the doctrine required for implementing strategic aerial warfare was available. In 1930 the Army had only 51 bombers, 131 attack planes, 309 pursuits, and a complement of observation, cargo, and training planes.[6]

Undaunted, the airmen set out to build their vision. Low appropriations, mishandling by the Army, and the controversies involving Mitchell had strengthened their cohesiveness, sharpened their professionalism, and given them the courage of revolutionaries. They seem not to have allowed themselves to doubt.

All of Walker's prior experience with bombers came into focus when he was a student at the Air Corps Tactical School (ACTS). Under instructors that included Maj Courtney H. Hodges and Capts George Kenney and Robert Olds, classroom lectures deviated from the operational record to enter the realm of the theoretical—what might happen *if.* A new course called "The Air Force" took shape, its purpose to integrate the functions of pursuit, bombardment, observation, and attack aviation into a single force.

Part of the ACTS curriculum included participation in maneuvers, and the one in which Walker participated as a student in 1929 proved important. ACTS personnel drew up a plan for an air phase prior to the ground exercise of the V Corps area maneuvers, held in Ohio in May. As the maneuver unfolded, ACTS personnel, serving as air umpires and as aviation staff officers of the ground units, not only dominated the air phase but made a significant impact on the ground phase through close cooperation of air and ground. The impact was profound, particularly for the bombardment arm. Maj Walter H. Frank, ACTS assistant commandant and chief umpire for the maneuver, reported afterward: "There is considerable doubt among the umpires as to the ability of any air organization to stop a well organized, well flown air force attack." Because pursuit had found it hard to find and intercept the bombers, Frank concluded, "a well planned air force attack is going to be successful most of the time."

The 1929 maneuver was a valuable experience for the school, since the practical test supported the directions they were pursuing in tactics, technique, and doctrine. The conclusion reached by Frank also made a profound impression on Walker, who over the succeeding years advocated it so convincingly and repeatedly that it became in many respects his own.[7]

In the spring of 1929, a group from the school attended a demonstration of chemical warfare tactics at Edgewood Arsenal. Poison gas dispersed from planes was simulated by aniline dyes. Those who marched down the road wearing gas masks and white hooded coveralls found themselves thoroughly splotched with green before the day was over. Obviously, there were many facets to consider about this new aerial dimension of warfare.[8]

Walker identified most closely with bombardment, which was taught by Robert Olds. Olds had served as an aide to Mitchell, and he brought Mitchell's influence to the bombardment class.[9]

Olds found Walker a ready partner. They caught the ball which Mitchell, after his court-martial, had tossed back into the younger officer ranks. Both men embraced all that Mitchell stood for—the independent air mission, the dominance of bombardment, even the need for aggressive advocacy. With the hero influence of Mitchell reinforcing all his experience and conclusions to date, Walker moved into full stride. He had found his professional niche. It seemed obvious to him that the bomber would replace the fighter as the determining element in the Air Corps of the future. This became his cause, his identity. The strongest image he left behind is that of the impassioned advocate, the spokesman for bombardment in the ongoing evolution of air force.[10]

Walker was one of 24 officers, 15 of whom were from the Air Corps, who graduated from the school in June 1929. He was assigned to the teaching staff for the coming year, along with Robert Olds and Donald Wilson.

As though realizing that his time was limited, Walker threw himself full-bore into the shaping of his profession. One of his first tasks was to rewrite and update the bombardment text. A more sizable volume than its predecessors, it was sent to the Chief of the Air Corps for approval in March 1930. It came back with minor suggestions and was revised. Approved in December 1930, it was then printed as a text, bearing the date February 1931.[11] Meanwhile, Walker had summarized current bombardment tactics in a substantial article, "Driving Home the Bombardment Attack," which was published in the *Coast Artillery Journal* of October 1930. In an orderly and reasoned manner, he addressed how bombardment planes could employ echelonment and other formation techniques to achieve maximum firepower and protection under varying circumstances. Faith in the bomber's ability to protect itself in the air showed strong, but not total; he acknowledged that the support of friendly pursuit might be required in several circumstances.

His conclusion, however, was clear. "It is generally conceded, by those who are competent to judge, that an air attack well launched is most difficult to stop."[12]

The conviction that pursuit could not stop a bombardment attack had been growing for several years. The problem was seen primarily in terms of interception; that is, defending pursuit simply would not be able to locate and destroy incoming bombers in time to prevent the bombing attack. The implications were profound. If the bombardment attack was always going to get through, then pursuit could have only a minimal effect on operations. In contrast, bombardment could take offensive action on the ground and simultaneously provide its own defense. The text for the Air Force course in 1930 incorporated this concept. "A strong hostile formation of bombardment or attack is likely to reach its objective before being intercepted and attacked by our pursuit. Even if attacked, unless by overwhelming numbers, it is likely to reach its objective."[13] Walker's article concluded with this statement: "The most efficacious method of stopping a bombardment attack would appear to be an offensive against the bombardment airdrome."[14] In other words, destroy an enemy's air force on the ground by bombing.

Much of Walker's 1931 text dealt with tactics, appropriate formations for particular circumstances, and administrative procedures for handling the bombardment units. His thinking, however, is obvious. He did not hesitate to state bluntly that bombardment aviation "is the basic arm of the Air Force."[15] He voiced no reservations about bombardment's ability to overcome defenses. Citing the experience of World War I, he noted, "Bombardment formations may suffer defeat at the hands of hostile pursuit, but with a properly constituted formation efficiently flown these defeats will be the exception rather than the rule. . . . Bombardment . . . will go through to the objective."[16] After discussing tactics for defending against antiaircraft, he concluded, "Bombardment personnel, indoctrinated with the will to reach and destroy the objective, will not be turned from their mission by the threatened or actual antiaircraft defenses of the enemy."[17]

Stopping short of calling bombardment decisive, Walker's text defined the primary function of bombardment as "to destroy objectives on land or sea by means of projectiles dropped from airplanes."[18] He considered this function sufficiently important that its control should be invested in General Headquarters (GHQ), the reasoning being that bombardment could operate far ahead of the front lines and against objectives outside the immediate concern of an army. Furthermore, since it was unlikely that there would be enough bombardment units to strike all desired objectives, choices would have to be made and could best be made by GHQ. Having included the basic premise of strategic bombardment, Walker went on to suggest the peg upon which the mature doctrine would later pivot: "There will probably be certain vital objectives comparatively limited in number which, if destroyed, will contribute most to the success of the combined arms of the Nation."[19]

Walker's text differed more in degree than in kind with the teachings of bombardment up to that time. Already established were the concepts that independent air operations could destroy an enemy's will to resist, and that these operations would be conducted by a bombardment plane. Not yet available, that bombardment plane would carry a heavy bomb load to a distant target, at an altitude and speed sufficient to escape enemy pursuit, and defend itself by machine guns and formation flight. The concept of targeting "vital parts" of an enemy's structure and destroying them by daylight precision attacks shows up in the text of 1926.

Walker's text did not intimate any reservation about bombardment's ability to accomplish its mission. Nor did it voice the doubt that had qualified earlier texts about the use of strategic airpower. The texts for 1924–27 noted that the strategic use of airpower against political centers was "prohibited by the laws of warfare," although they went on to note that political centers were apt to contain important military objectives, that they would likely be attacked by an enemy, and that reprisal against attack was justified.[20]

The moral issue of bombardment was an important one, and it received thoughtful analysis. Although deliberate attack on civilians and cities as a demoralizing measure was rejected, the degree to which industrial and political targets constitute a military objective was a more open issue. Just as warfare was undergoing dramatic change, with greater mechanization and more deadly weapons, so too was the concept of morality in warfare in a state of flux. Lacking diaries, letters, or other concrete evidence of Walker's conclusion, we can only speculate that he took part in these discussions and resolved to his satisfaction any moral conflicts that arose for him. He and his fellow officers believed that the bomber could wreak tremendous destruction, which would exercise a decisive effect on an enemy's ability or will to fight. They anticipated that bombardment would result in shorter and less costly wars, with fewer deaths than the ground warfare of 1914–18. The eventual doctrine which Walker so fervently advocated—precision bombardment on industrial targets—was designed to keep the application of airpower within the moral framework of the laws of war.

ACTS was in many respects an intellectual/professional stew, with all its components contributing to the flavor. One participant described it as a "good collective brain," an atmosphere "vibrant with development of new concepts of warfare" within which Walker was one of the stars.[21]

Certainly, Walker's thinking shows up in the text for "The Air Force" course in 1930 and 1931. That text positioned air as subordinate but vital to ground forces, unlikely to overcome an enemy alone. But it also leaned heavily toward bombardment as the primary arm of an air force and acknowledged a distinct strategic purpose for airpower. An air force is like an army, it stated, "a unit composed of several dissimilar components; whose objective may be immediate, contributory to a larger purpose; or, more often, ultimate—the complete destruction of the opposing force or of his will to continue hostilities."[22] As for fulfilling its mission, the text held that bombardment planes could defend themselves even against odds of two to one; the mission would go through.[23]

Not all airmen agreed. One of the students at ACTS during the 1930–31 year, Walker's first year as an instructor, was Capt Claire L. Chennault. As volatile in temperament and firm in conviction as Walker, he too had joined the Regular Army Air Service in 1920. While Walker had been acquiring experience and expertise in bombardment, Chennault had been doing the same in pursuit. Seeing his branch of the Air Corps being written off as obsolete, he rebelled. The stage was set for a debate of such heat and extended duration that it became legendary within the Air Corps.

Their six years at Langley Field were good ones for the Walkers. They lived in base quarters known as "Lighter than Air," with Muir and Florence Fairchild just across the backyard. In the summer of 1926, Lt William K. Andrews and his wife Mamie Lee moved into the other side of the Fairchild duplex. Andrews, who had served in France during the war, was back in service for a short enlistment. He was assigned to the 11th Bombardment Squadron, of which Walker was at that time commanding officer. Before long, the two couples had formed a strong and constant friendship. Marguerite was pregnant with her first child; Mamie Lee soon found out that she too was pregnant. The two women shared their experience with much happiness, among other things gorging themselves on the wonderful refreshments served at the weekly meetings of the "Pregnant Girl's Club."[24] The Walkers' first child, Kenneth N. Jr., was born 18 February 1927. His proud father made certain his new son had appropriate toys—a pedal-driven airplane and a pilot's coverall and helmet.[25]

Langley's officers and their families formed a tight, mutually supporting society within which the Walker household was one of the more lively and social ones. Walker's body had grown lean as he matured, so that he seemed tall. He was supple and athletic—he played a lot of squash—and was known as a wonderful dancer. His favorite arrangement (Paul Whiteman) of "Rhapsody in Blue" was often on the record player. The women found him handsome despite a prominent nose that they tactfully conceded was not his best feature. He dressed with a sense of style; his smart clothes set off his good looks.[26]

"Langley!" Gen Haywood S. Hansell reminisced with Marguerite years later: "What fun we had when we were all so young." He recalled "dances at the Officer's Club to the strains of music provided by the struggling efforts of an orchestra from the base band, with the men all wearing boots and spurs! And the occasional—and more expensive—dances at the Chamberlain at Fort Monroe." Hansell marveled that they got along without air-conditioning: "We just didn't know how much we suffered from the heat."[27]

Custom required much of an officer's wife in the role-segregated society of the base, and Marguerite was a gracious and talented hostess, a "quiet and lovely lady." Ken provided little or no help with the chores of entertaining, but together the Walkers became one of Langley's institutions—the dashing officer husband, intelligent and intensely involved in his career, and the beautiful and charming wife. These were exciting times, and the officers at Langley—particularly those connected with the Air Corps Tactical School—were the Air Corps' activists. Joining them in 1930 were 1st Lt Laurence S. "Larry" Kuter and his wife Ethel, who perceived Walker as good-looking and self-contained, but "totally dedicated to his work." The dominating facet of his personality, in her opinion, was his near-total involvement with himself and his ideas. "Single-minded and high-strung," confirmed Hansell, but like others he pointed out an opposite side to Walker's nature: he was affable and relaxed when not at work. Walker's primary image, however, was that of a man obsessed by what he was doing.[28]

Walker was not alone in his enthusiasm. "The men flew all day and all night," Mamie Lee Andrews recalled, although the aviation talk did not include the women. At the frequent social occasions, "the men went off and talked flying and the women talked children and food." To the extent that the women got to know the men, it was through their wives, for all the wives belonged to the Officers Wives Club. ("There was no 'joining,'" Ethel Kuter said; "you just were!") Seeing him more closely than some, Mamie Lee Andrews' perception of Walker was that he was "so very tense, so active mentally, as well as so athletic." She marveled that he stayed lean despite his hearty appetite; he walked home for lunch every day and Marguerite always had dessert to satisfy his sweet tooth.[29]

Walker's tense impatience sometimes led to grief. On 28 April 1929, shortly before graduation from the ACTS, he flew an AT-5 on a cross-country navigation mission to Columbia, South Carolina. On the way back, he ran into heavy rainstorms and tried to avoid them by going east of his course. When the storms worsened, he put down in a field. An hour and a half later, with the weather improved, he tried to resume his flight. The field was not only wet, but plowed; he had to take off with the furrows, even though this meant a cross-wind takeoff. For 150 yards all went well. Then he hit a rough spot in the field, left the ground for a few feet, and touched down about 10 yards farther on. The right wheel plowed into a mound of soft dirt and the plane nosed over on its back to suffer a bent propeller and other damage. Walker suffered the pang of pride at being advised that he "should have spent more time in finding a more suitable field from which to take off." It is unlikely that he worried much about it. The incident serves, however, to illustrate the status of aviation. Walker and his colleagues could spend hours of theoretical planning for using airpower to defeat the nation's enemies, but at this point they had no plane or supporting facilities capable of anything close to such a task.[30]

During the summer of 1931, the Air Corps Tactical School moved from Langley to Maxwell Air Force Base, Montgomery, Alabama. That summer, the Walkers spent a delightful and restful month's leave with Bill and Mamie Lee Andrews at a rented summer cottage at Biloxi, Mississippi. Once they moved to Montgomery, however, life resumed the frenetic pace it had begun to acquire at Langley Field. Until August 1933, Walker served as the ACTS instructor in bombardment aviation. For everyone in the service, those were years of economic depression and low appropriations; there was no money to push technology ahead. Nor did the air advocates make much progress vis-à-vis the US Army over what they saw as a fundamental issue: would air be administratively subordinated to ground officers or have its separate organization? Nevertheless, during Walker's years as one of the driving forces at the ACTS, a doctrine for strategic air warfare took shape, essentially intact, ready for later refinement, acceptance, and implementation. The stormiest years were 1931–33.

At the beginning of the 1931 year, Hume Peabody, the school's assistant commandant and instructor for "The Air Force" course, told the instructors that each was free to teach what he wanted to. "Then with the ideas we get from the present students, we are going to hit a happy medium in here where we can all work together."[31] What followed was an extraordinary, extended debate. Walker, as instructor in bombardment, held a pivotal position. Those who observed him during these years invariably mention his intensity, his chain smoking, his habit of pounding furiously away at his typewriter in a posture of frustration and near frenzy. His position grew steadily more clear and more firm: a strategic role for airpower, implementation by daylight precision attacks on critical industrial targets, and offensive bombardment rather than pursuit as a deterrent or defensive force.

This approach evolved through the efforts of a number of officers, but Walker, Harold L. George, Robert M. Webster, and Donald Wilson were outstanding among them. Each came to the same conclusions independently and at about the same time. Mass raids on civilian populations were not morally acceptable, hence precision raids on industrial targets. For precision, one needed daylight; and daylight raids would be opposed by antiaircraft and hostile pursuit. Bombing must therefore be done from high altitude to avoid antiaircraft; defensive guns and formation flight must protect from pursuit. Wilson's primary contribution was target selection, which made it feasible, in view of the size of force that could be applied, to think in terms of paralyzing an enemy by knocking out specific keys to important industrial systems.[32] In "American Air Power in World War II," May 1968,[33] Hansell credited a wider group of officers for the doctrinal evolution. In addition to Walker, George, Webster, and Wilson, he named W. S. Frank, George C. Kenney, Joseph McNarney, Muir S. Fairchild, Hoyt S. Vandenberg, Laurence S. Kuter, and Gordon Saville, "to name only a few."[34]

Above all, Walker argued that the bomber would get through. His conviction was ardent, shored up with passion, and held with the surety of one who has resolved to his own satisfaction any doubts or reservations. The bomber's ability to get through was the pivot for the total debate, for otherwise there would be no means whereby independent air action could affect an enemy's will. In his writing, his lectures, and his ongoing debates with the doubters, Walker insisted again and again that a determined air attack, once launched, would be impossible to stop. In his text, he stated that the bombardment airplane "must have the ability to protect itself from hostile air attacks"; he discussed the placement of guns, the importance of high ceiling and rapid climb, and even a reduction in noise to make it easier for bombardment to slip past enemy observation.[35] He maintained that close formation with mutually supporting fire would provide defense against hostile pursuit while speed, maneuverability, altitude, clouds, and darkness would provide defense against antiaircraft.[36] He saw the bomber as the basic aircraft type; all other branches of the Air Corps should be built around bombardment with the purpose of ensuring the success of the bombing attack.[37]

Pursuit figured less and less in Walker's thinking, but Claire Chennault wasn't buying. For intensity, stubbornness, and vehemence, he matched Walker round for round, often with little help from others. Walker had broader support and two formidable stand-ins: George and Wilson. Those three maintained that the bombers would get through, that pursuit could not intercept quickly and accurately enough to affect the battle, and that armed bombers flying in defensive formation could conduct offensive strikes without pursuit escort for protection.

"No way," replied Chennault, who insisted that an offensive bombardment mission as proposed would be thwarted by enemy pursuit, which would shoot down the bombers. Improved planes plus a system of warning nets, he contended, would give pursuit the edge it needed for interception. With the interception problem solved, fighter planes would continue to be necessary for achieving and maintaining control of the air. Some of the men in bombardment quietly agreed, saying they would take all the fighter protection they could get. George Kenney predicted that the bombers would hit flak "so heavy you could walk on it." Walker and the other bomber enthusiasts were unmoved—"hidebound," according to Peabody.[38] Both Walker and Chennault argued largely on faith—faith that planes could and would be developed to implement the aerial warfare they visualized. Here Walker had the edge, for the available technology of the early thirties favored the bombardment class.

The personal mannerisms of the participants colored the scene. The grim-faced Chennault, known for blunt words spoken in a soft southern drawl, had a habit of snorting. Walker wore heavy glasses "thick as coke bottle bottoms" and was constantly looking down over them or pushing them back. Chennault smoked, but Walker smoked even more heavily and was described as more emotional than Chennault. Observers used the word "rabid" to describe both, even suggesting, "If they had had tomahawks, they would have scalped one another." Some officers enjoyed hearing the verbal battles and even provoked them intentionally; others claimed that the arguments, plus the exercises staged to test their ideas, provoked ulcers.

No holds were barred. "You got the pure gospel according to either," recalled Gordon Saville. Chennault seemed to be "always putting us on the line, saying, 'Let's go out and try out for real.'" Astounded that the bomber advocates actually believed the things they were saying, Saville finally went to

Chennault and said, "Look, if you ever succeed in getting three bombers up, armed, and three fighters, I'll be one of the fighters. We'll go and shoot them, live bullets."[39] Howard C. Davidson, an ACTS student during this period, conceded that in debate Walker gave nothing, but even so Davidson considered him "just about the best instructor" at the school, a friendly guy who was well-liked.[40] He displayed a keen and biting sense of humor; he pursued his recreation as vigorously as his work. He drank moderately but with relish, buying white lightning from the locals and aging it himself because of prohibition. An hour's horseback riding every morning was still a requirement at the ACTS, much to the irritation of many, but Walker seemed actually to enjoy it. He also enjoyed singing cowboy songs while accompanying himself on the guitar. Hansell, who recalled that Walker's voice quality matched that of his beverage bar, attributed all this to his New Mexico and Colorado background.[41]

Walker also continued to take pride in his dress. In early 1932, when the Air Corps command rejected a proposed new uniform that did away with boots, Walker took time to write a semispoofing letter pointing out that "to put on a pair of well fitting boots is a problem, to remove them is an adventure." Since boots implied riding breeches, and if breeches fitted properly the knees got ripped out in climbing in and out of airplanes, Walker proposed they wear golf breeches and boots with a flexible top, such as those worn in Sweden. "Don't give up the fight on an Air Corps uniform," Walker told his correspondent. "You are sure to drive home eventually the idea that an Air Corps officer doesn't have to be dressed as though he were leading a cavalry charge in order to present a snappy appearance." The snappy appearance being important to him, he invested in a pair of specially fitted, handmade Peal boots.[42]

Despite humor and diversions, the atmosphere at the ACTS was contentious and tense. Walker did not shirk it. A 1933 letter to fellow officer Carl A. "Tooey" Spaatz, however, indicates he was worried about the internal rivalries and prejudices building within the Air Corps because of "considering ourselves as bombardiers or pursuiters." He suggested that training be changed so every officer could fly any type of plane and learn the minor tactics of each branch. "I feel that any step that we may take to eliminate internal prejudice is worthwhile and practical." Ideally, he wanted officers "thinking in terms of air force as well as in one particular branch." Where, he asked, is the place for "this fetish of overspecialization?"[43]

Overspecialization or not, the name Walker was almost synonymous with bombardment. Then and later, he was the individual within the pre-World War II Air Corps most closely identified with the credo, "A well-organized, well-planned, and well-flown air force attack will constitute an offensive that cannot be stopped."[44] Some even attributed to him the introduction of the word "bombardier," which began to replace "bomber" in Walker's writings.[45]

Beyond doubt, Walker believed in his profession and felt positive about his place within it. "The Air Corps has achieved a position for which it need not apologize," he wrote Spaatz in 1932. "It no longer has to sell itself to the rest of the Army." Asked to critique an antiaircraft exercise Spaatz had conducted at Fort MacArthur, he urged closer cooperation between air and antiaircraft so that the air's own tactics might be improved. "The spirit of contests," he said, "should be eliminated" from exercises. He wanted to show that the Air Corps "is desirous of playing ball as well as the fact that the antiaircraft is not as hot as they think it is." With no indication that he realized the same could be applied to him, he mentioned the need to "quiet some of these wild-eyed enthusiasts who continually state opinions as though they were facts." He in turn struck the open and receptive posture of the secure: "We [at the ACTS] are very anxious to receive all suggestions and recommendations concerning the improvement of tactics here." Pleased that the units at March Field had used tactics developed at the ACTS for their exercise, he noted that practical experience "is always limited." He hoped units in the field "will take the school tactics, use them and criticize them constructively, and with them as a starting point, develop superior tactics."[46]

Despite Walker's apparent receptivity to ideas for improving tactics, his mind was closed concerning the basic tenets of his faith in bombardment. He had the inner conviction and moral courage to fight for that faith, and even though we might wish he had been more receptive to those who saw its weaknesses, had he been more flexible he might have achieved less rather than more. The spear that pierces deep has a straight, undeviating shaft.

These pre-World War II aviation officers faced a complex scenario wherein experience was thin, money scarce, and encouragement scant. They based their arguments on theory, speculation, and faith. It was possible to hold opposing positions with sincerity, as did Chennault and Walker. Exercises and maneuvers helped, but of necessity were contrived; opposing sides could draw opposite conclusions from their results. Walker helped plan the Fort Knox Maneuvers of 1933, a major purpose of which was to test the use of a distant intelligence net such as Chennault advocated. The final report, which Walker helped prepare, stated that the warning net operated "very satisfactorily and efficiently," and Chennault believed this proved all he had said about the vulnerability of bombardment to pursuit. There is no evidence, however, that Walker or other bombardment advocates changed their thinking. To the contrary, other maneuvers held that year prompted them to conclude that "no known agency [can] frustrate the accomplishment of a bombardment mission."[47]

Apart from maneuvers and exercises, there was little else to go on. Walker did some work with probability theory, analyzing results of bombing experiments to calculate error.[48] Technological evidence was more concrete and in the end became decisive. Bombers as a class were advancing more rapidly and showing more promise for future development than were fighters. Had there been money to pour into plane development, the picture might have been different. Funds, however, were limited. Since the most essential component of the strategic air doctrine was a long-range bomber, the Air Corps elected to believe that pursuit did not have the same technological potential that bombardment enjoyed. The limited funds for design and development went toward what became the first truly modern bomber, the B-17.

Walker and his contemporaries overestimated what bombardment could accomplish, underestimated the effects of friction and enemy defense, and failed to develop pursuit or to integrate airpower with other forms of warfare. What they did achieve, however, was to create an air arm with a foundation sufficiently strong and an officer corps sufficiently mature that air forces could adapt to the realities of warfare and play a major military and strategic role between 1941 and 1945. The foundation of that air force was, as Walker insisted it would be, bombardment.[49]

There is no evidence that Walker at any time held serious reservations about the evolving aerial doctrine. An impassioned advocate of his cause, he had adopted some of Mitchell's tactics, including that of denying the integrity of the equally dedicated opposition.[50] By 1935, when the Norden bombsight was introduced and the B-17 was successfully flown, bombardment was in ascendancy and pursuit had become only a minor part of the air arm.

By that time, Walker was no longer at the ACTS. In August 1933, he entered the Command and General Staff School (CGSS) at Fort Leavenworth, Kansas. At Walker's encouragement, George stayed to teach bombardment and, later, to direct the Department of Air Tactics and Strategy. At the ACTS, the refinement of doctrine continued, more confidently after Wilson began working on the concept of the industrial fabric. Responding to the "certain vital objectives" Walker had suggested in his 1931 text, targeting was based on the theory that a nation's industrial capacity could be neutralized by eliminating key elements of it—ball bearings, for instance—without which a number of other industries could not function. Target selection, coupled with high-altitude daylight precision bombing, gave the Air Corps a philosophy around which to plan its future role. The bombardment text for 1933–34 included the statement, "bombardment aviation, properly employed, can shatter a nation's will to resist; it can destroy the economical and industrial structures which made possible the very existence of modern civilization."[51]

Notes

1. "Pilots Showed Speed of Bombing Planes," *New York Times*, 11 September 1926, 17.

2. Maj Gen Hugh J. Knerr, interview from Walker Medal of Honor Scrapbook, 2 October 1971, Office of Air Force History (OAFH).

3. Thomas H. Greer, *The Development of Air Doctrine in the Air Arm, 1917–1941*, USAF Historical Study 89 (Maxwell AFB, Ala.: USAF Historical Division, Air University, 1955), 57.

4. DeWitt S. Copp, *A Few Great Captains: The Men and Events That Shaped the Development of U.S. Airpower* (Garden City, N.Y.: Doubleday, 1980), 272, 274–75.

5. Robert T. Finney, *History of the Air Corps Tactical School, 1920–40*, USAF Historical Study 100 (Maxwell AFB, Ala.: USAF Historical Division, Air University, 1955); Robert Frank Futrell, *Ideas, Concepts, Doctrine: Basic Thinking in the United States Air Force, 1907–1960* (1971; reprint, Maxwell AFB, Ala: Air University Press, 1989); and Edward O. Purtee, *History of the Army Air Service, 1907–1926* (Wright-Patterson AFB, Ohio: Historical Office, 1948).

6. Clayton Bissell, "Brief History of the Air Corps and Its Late Developments," call no. 358.40973, 1 January 1927, U582b, Air University Library, Maxwell AFB, Ala., USAFHRC.

7. Futrell, 33. See also Finney.

8. J. D. Barker, *History of the Air Corps Tactical School*, 245.01-B USAFHRC.

9. Finney, 27. Quotes Kuter in identifying Walker, also, as an aide to Mitchell. But Mitchell's biographer, Alfred Hurley, recalls no mention of Walker among Mitchell's papers (Hurley to author, 18 February 1991), and Walker's 201 File contains no duty with Mitchell. No evidence that Walker worked directly with Mitchell has been found.

10. Finney, 27; Greer, 56.

11. Air Corps Tactical School (ACTS), Bombardment Textbooks, 1924–1925, 1926, 1930, February 1931, October 1933, 30 July 1935, and 1 January 1938, 248.101-9, USAFHRC.

12. Ibid., 340.

13. *The Air Force*, text, April 1930, 248.101-1 USAFHRC, 73.

14. Ibid., 340.

15. Ibid., 69.

16. Ibid., 93.

17. Ibid., 99.

18. Ibid., 16.

19. Ibid., 69.

20. *Bombardment Aviation*, texts, 1924–1925, 1926–1927, 248.101-9 USAFHRC, 68 in 1924–1925, 63-64 in 1926-1927.

21. Gen Laurence S. Kuter, oral history interview with Hugh N. Ahmann and Thomas A. Sturm, 30 September–3 October 1974, K239.0512-810, 161, 163. Laurence S. Kuter, unpublished manuscript, US Air Force Academy Library Special Collections, 74–75.

22. *The Air Force*, text, *1930, 1931*, 248.101-1, USAFHRC, 30–31 in 1930 edition, 4; 100–101 in 1931 edition.

23. *The Air Force,* text, 1931, 88–89.

24. Mamie Lee Andrews, telephone interview with author, 28 August 1990.

25. Kenneth N. Walker Jr., Berkeley, Ontario, interview with author, 8 March 1991.

26. Mamie Lee Andrews, Chatham, Mass., interview with author.

27. Haywood S. Hansell to Marguerite Walker, letter, 16 July 1986.

28. Quotations from telephone interview with Ethel Kuter, 29 January 1991. Other details from telephone interview with Mamie Lee Andrews, 28 August 1990. Haywood S. Hansell, "Brig. Gen. Kenneth N. Walker: Prophet of Strategic Airpower," *Air Force Magazine,* November 1978, 92–94.

29. Telephone interview with Mamie Lee Andrews, 28 August 1990; and Ethel Kuter, telephone interview with author, 29 January 1991.

30. Accident Report, 200.3912-1, USAFHRC.

31. Hume Peabody, oral history interview with Hugh N. Ahmann and Thomas A. Sturm, 30 September–3 October 1974, K239.0512-810, USAFHRC, 305.

32. Greer, 57–58.

33. Haywood S. Hansell Jr., "American Air Power in World War II," manuscript, K112.3-2, USAFHRC.

34. Donald Wilson, "Origin of A Theory for Air Strategy," *Aerospace Historian* 18, no. 1 (March 1971): 19.

35. *Aerospace Historian,* 25–26.

36. Ibid., 70.

37. Kenneth N. Walker, "Bombardment Aviation: Bulwark of National Defense," *US Air Services,* August 1933, 15–19; Greer, 56.

38. Martha Byrd, *Chennault: Giving Wings to the Tiger* (Tuscaloosa, Ala.: University of Alabama Press, 1987), 39–40, 45–46; Partridge, interview, 217; and Peabody, interview, 307.

39. Maj Gen Howard C. Davidson, oral history interview with Hugh N. Ahmann and Thomas A. Sturm, 5–8 December 1974, K239.0512-817 USAFHRC, 362, 369, 385; Maj Gen Gordon P. Saville, oral history interview with Thomas A. Sturm, K239.0512-1322 USAFHRC, 35, 41. Quote about glasses from Douglas C. Shelton to Douglas P. Walker, 18 January 1991.

40. Davidson, interview, 385.

41. Unedited draft of Hansell's article, "Brig. Gen. Kenneth N. Walker," in Hansell Papers, Box 2, US Air Force Academy Library, Special Collections.

42. Kenneth N. Walker Sr., to Major Weaver in Washington, 5 February 1932. Photocopy sent to author by Douglas P. Walker.

43. Quoted in Copp, 260.

44. Futrell dedicated his comprehensive *Ideas, Concepts, Doctrine* to Walker as the one who taught that creed. Haywood S. Hansell tied the slogan even more closely to Walker. See Hansell, "Brig. Gen. Walker," 92. In a letter to Marguerite Walker, 16 July 1986, Hansell used the term "coined the slogan."

45. Douglas C. Shelton to Douglas P. Walker, letter, 18 January 1991. In his bombardment text, Walker used the term "bomber"; but in a letter to Carl A. Spaatz (cited in Copp, 260), he used "bombardier."

46. Walker to Carl A. Spaatz, letter, 3 September 1932, and accompanying "A Study Concerning Air Corps - Antiaircraft Exercises Conducted on the Night of 15 April '32," 248.2122-2A, USAFHRC.

47. Joint Antiaircraft-Air Corps Exercises, 15–27 May 1933, Report no. 18.343.1B, in Benjamin D. Foulois with C. V. Glines, *From the Wright Brothers to the Astronauts: The Memoirs of Benjamin D. Foulois* (New York: McGraw-Hill, 1968), 229.

48. Haywood S. Hansell Jr., *Strategic Air War against Japan* (unpublished manuscript, Maxwell AFB, Ala.: Air War College, Airpower Research Institute, 1980).

49. For a similar conclusion, see Thomas A. Fabyanic, "A Critique of United States Air War Planning, 1941–1944" (PhD diss., Saint Louis University, 1973), 7, 45–48.

50. Alfred Francis Hurley, *Billy Mitchell: Crusader for Air Power* (New York: Franklin Watts, 1964), 139.

51. Donald Wilson, "Origin of a Theory for Air Strategy," *Aerospace Historian* 18, no. 1 (March 1971): E 19; and United States Army Air Corps, *Bombardment Aviation, 1933–34*, 248.101.9, USAFHRC, 81.

Chapter 3

More Schooling and Command

Assignment to Command and General Staff School (CGSS) took Walker out of doctrinal development, but it placed him in the middle of the Air Corps struggle for control of its own organization. About the time he went to CGSS, Walker published another professional article that presented an adequate air force as a deterrent to enemy aggression. "Bombardment aviation has confined wars to continents!" he declared, for "No enemy would consider launching an invasion . . . if he were convinced that we were in possession of a bombardment force capable of destroying [vital enemy establishments]."

By this time, he had dismissed pursuit; "whenever we speak in terms of 'air force' we are thinking of bombardment aviation." An adequate air force, which could destroy both airplanes and their facilities on the ground, "would prohibit air operations against us." He restated his major thesis, the pivotal point on which Air Corps doctrine was evolving: "a determined air attack, once launched, is most difficult, if not impossible to stop when directed against land objectives."

At the end, he included a point about organization. Calling for the country to build an adequate air force for defense, he said this could be done only "when the importance of an air force to our national security is appreciated fully—not in terms, for example, of a mere adjunct to our ground forces, such as cavalry or field artillery and designed *only* to further the infantry mission—but as a force with a distinct mission, of importance co-equal to that of the Army and the Navy."[1]

The article was published in August 1933, the same month Walker entered CGSS. For the most part, airmen considered the staff college wasted time. There was little in the curriculum pertaining to air, and as Donald Wilson put it, "it was just silly to send air officers to . . . learn the minutia of ground officers' duties, which we did."[2] Attendance at CGSS was considered essential, however, if an officer expected to advance in rank. At this time, Walker was still a first lieutenant—there were jokes about his being the most senior first lieutenant in the Air Corps[3]—for although he made first lieutenant when he entered the Regular Army on 1 July 1920, he reverted to second lieutenant on 15 December 1922 and did not regain first lieutenant rank until 24 July 1924. Perhaps CGSS pried him loose, for he made captain shortly after graduation in June 1935.

Between the world wars, the issue of Air Corps organization ran parallel with the search for doctrine. Conflict had begun with the Army Reorganization Act of 1920, which placed air units under the control of ground commanders as an integral part of ground forces. Prior to his death in 1936, the outspoken Mitchell kept controversy constantly before the public, always reinforcing the airmen's position that the air arm should operate as a separate striking force and air strategy should be handled by those most interested and best qualified to plan and direct it—air officers rather than ground officers. There was an element of turf war involved in this controversy for control, and the controversy had a close relationship with the evolution of strategic air doctrine. If air's contribution was primarily observation and tactical support of ground forces, as the Army contended and limited experience indicated, the case for a separate department was less compelling. Although it can be argued that airmen needed the independent organization in order to implement their doctrine of independent operations, it can also be argued that they developed the doctrine to justify an independent organization, the latter being desired for their own need for professional respectability.[4]

Certainly, the air arm attracted its share of bright, creative, dynamic individuals. Accorded less than full weight within the Army, they took their case public. The press gave airmen and air activities good coverage because they were highly visible and dramatic—they made "good copy." Press coverage in turn made the Air Corps case a public issue in which Congress became involved. The result was a succession of official hearings, boards, and commissions, assembled for debate on the issues. In late 1934, Walker was invited to testify before one of them—the new Federal Aviation Commission, which was charged by President Franklin D. Roosevelt with recommending to Congress a broad policy for all phases of aviation. The board's head was Clark Howell, editor of the *Atlanta Constitution*; he promised to probe the issues of military organization with an open mind. That was unwelcome news to the Army. The recent Baker Board had recommended that the existing organization stand; it saw air as a homogenous part of the Army and consequently rightly subordinated to the general staff. The Baker Board did, however, recommend a General Headquarters Air Force that would bring some coordination to Air Corps action. The secretary of war endorsed the Baker Report, and the general staff advised the officers now summoned to testify before the Howell Commission to conform to official policy.

"Official policy" was not necessarily Walker's forte. On a cross-country flight about this time, he stopped in Quincy, Illinois, to see Bob Pearson. He apparently sounded off to his friend, for afterward Pearson remembered that Walker marveled at the resistance of the military mind to new ideas. Pearson retained a strong impression of Walker's independent thinking and admired him for not climbing on the official bandwagon.[5]

So it seems to have been an unintimidated Walker who began gathering his arguments and building his position for the hearing. Several of his close friends were also summoned—Olds and Wilson from CGSS, along with George from the ACTS. They protested the War Department's attitude and asked to be officially designated as witnesses, to be issued official orders to attend, and to be given military transport. They won the transport, but their attendance would be "entirely voluntary" and they were warned not to give personal opinions without clearly identifying them as such. The men assumed their careers were on the line. They met and talked it over. With courage to match their convictions, they decided to be forthcoming and present the issues as they saw them. They did so, exercising admirable emotional restraint and professional prowess. At the conclusion, even Brig Gen C. E. Kilbourne, Assistant Chief of Staff, War Plans Division, called theirs a "constructive presentation."

In their testimony, George emphasized air as a method of waging war rather than a weapon comparable to a rifle. He projected that future wars would be fought with the air forces in existence when war broke out, hence the importance of an air force in being. Since air would be in action for some time before ground forces were, he said it needed a separate organization. Olds reinforced George's points, saying that what we maintain during peace is what we have to fight with when war begins. Both Army and Navy need aviation as a part of their forces, he said, but he called for a national air force in addition.

Wilson noted the international climate and observed that this was no time "to pin our faith on brotherly love." We should give a potential enemy credit, he said, and assume he will do that which is most detrimental to us—attack our industrial areas. The primary issue, in his view, was the national defense.

Walker also stressed the national defense and said the security of the nation demanded an adequate air defense. "National Defense is not the responsibility *alone*, of an Army, a Navy, or yet of an Air Force. It is the mission of the combined forces, in which each must play its part." The air mission, he said, was "to destroy and disrupt an enemy's means of waging war against us." He then spent a good portion of his testimony elaborating on this aerial mission and showing that neither Army nor Navy was suitable to direct it. He advocated a Department of National Defense with a Superior General Staff, but he acknowledged that such a major reorganization was not timely. He therefore endorsed George's proposal for a separate air organization under the War Department as "a definite and progressive step that certainly can be taken at this time." He believed any problems related thereto could be solved. "Unless we create an adequate and separate Air Force," he concluded, "this next war will begin in the air and end in the mud—in the mud and debris of the demolished industries that have brought us to our knees."

After hearing 191 witnesses, the commission concluded: "The history of American aviation has not been conspicuous for unanimity of opinion." It declined to muddy the waters any further. Fearing that three services would be no easier to coordinate than two, it did not recommend a separate air organization. It did, however, state that "there is ample reason to believe that aircraft have now passed far beyond their former position as useful auxiliaries, and must in the future be considered and utilized as an important means of exerting directly the will of the Commander in Chief." Walker and his associates could feel justifiable pride in the Air Corps they had shaped thus far. They had pushed and pulled their evolving branch of the service along poorly defined and uphill routes, but they could see the summit by 1935. While acknowledging that the Air Corps had "serious internal problems," the commission expressed "only praise" for the professional ability of its personnel.[6]

Walker's personal life began to fall apart during the years at Maxwell Field and Leavenworth. He left little personal evidence of his private thoughts and feelings; the reactions and memories of others form an outline that suggests—but does not reveal—his inner pain or motivation. Friends and family agree that he adored his mother, that her hard work to provide for him probably accounted for some of his tremendous drive, and that growing up without a father had left hurts and insecurities within him that showed themselves in subtle ways. There was, of course, a dark side to his nature. Despite their close friendship, Mamie Lee and Bill Andrews were sometimes disturbed by Walker's rowdy practical jokes, for they seemed to convey a hint of coldness and cruelty. Mamie Lee called him "the worst kind of a tease," the sort who might enjoy scaring a child. Bill vividly remembered a time when they had been flying together and slept that night in a tent. While he was asleep, Walker put a pillow over Andrews's face so that he woke up in the panic of suffocation. Walker seemed to enjoy the prank thoroughly but Andrews did not.[7] Hansell, also a good friend, described Walker as having "a brash sort of humor that went with a brash and often abrasive assertiveness."[8]

Walker also exhibited a crude racism, doubtless reflective of the times in which he lived but also suggesting those personal insecurities that encourage the domination or diminution of others. When stationed in the Philippines, the Walkers and Sheltons sometimes played an insensitive game with the Filipino students, dispersing groups by kicking butts. By kicking rather than hitting, the Americans reinforced the racist understanding that assigned the Filipinos to a lower social class. On another occasion in the late twenties, probably on the cross-country flight of Knerr's 2d Bomb Group, Walker and some others were, as Douglas Shelton remembered it, "flying all over the country" in "a bunch of bombers" and landed at Kelly Field, Texas. Walker spent the night with the Sheltons at Fort Sam Houston, Texas. It was the last time young Douglas saw his "Uncle Ken," and a small incident that took place stuck in his mind. "The black maid's kid and I were close friends and Uncle Ken was irked that Harve and Gin allowed him to call me Doug instead of Mister Doug."[9]

Walker, like his adversary Chennault, showed a defiant defensiveness that can be at least partially explained by career circumstances. Neither had attended West Point or served overseas in World War I, the two unofficial prerequisites for advancement and influence. The entire Air Corps of the twenties and thirties received little professional recognition or encouragement within the Army; its officers were largely outside that inner circle of the old West Point Army. Circumstances forced Air Corps officers into combative relationships for self-preservation. Chennault's chip on the shoulder showed up in brusque burliness and studied indifference to Army protocol; he said what he thought and went his own way, muttering about "the foggy-brained brass" who forced every airman to become a "belligerent crusader."[10]

Walker took the opposite course, dressing with style, entertaining with flair, and playing the social game, although he too could talk without pulling the punches when called upon. Both men had compulsive working habits, both were described at times as frenzied, or out of control. Both of them had health problems that were worsened, if not precipitated, by their self-imposed professional pressure. Chennault retired due to poor health in 1937. Sometime before that, Walker sought therapy at a warm springs for treatment of arthritis, usually considered a stress-related disease. The therapy seems to have put the arthritis in remission, but during the course of his therapy Walker met and became involved with another woman.

Walker apparently hoped to keep both his marriage and his affair. The Walkers' second son, Douglas, had been born in January 1933, and Walker often stated that he never wanted his sons to have the hard childhood he himself had led. Now, however, he was precipitating exactly that sort of difficulty. He discussed the problem with Marguerite, doubtless appealing to her concern for their children. She understandably found the situation deeply hurtful. More a lady than a fighter, and concerned for their family, she tried to accept his claim that his mistress provided him with things she could not give him. Oddly enough, among these was a Doberman pinscher, an unfriendly dog which Ken Jr. remembered as showing no interest in anyone at all but his father.

Young Ken felt the effects of the mounting tension between his parents. His relationship with his father became more formal and serious. If he failed to walk in a military manner, he would hear a sharp, "Straighten up there, soldier!" When he reported being attacked on the way home from school by bullies who held him down and tickled him, he was told to learn to fight back. Concluding that he was expected to take on a military relationship to his father, with father as the senior officer, he remembers that period of his life as emotionally grim, devoid of feelings of warmth or comfort from his parents.[11]

In the fall of 1933, the family moved to Leavenworth. Young Ken was given a happy memory of his father when they went flying together in a two-seater, but the flight also provided the boy a moment of sheer terror when Walker dipped a wing so he could point out their house.

While at Leavenworth, the Walkers decided to end their marriage. Friends were hurt, shaken, and horrified. One general stormed home to tell his wife, "My God, you've got to do something!" Agreeing that it was dreadful, the wife professed to know of nothing she could do. "Don't give me any of that," the general shouted back. "You're always taking care of things. The Walkers are an institution! You can't just let an institution break up!"

But break up it did. Divorces were easy to obtain in Arkansas, where Harvey and Virginia "Gin" Shelton now lived in Conway. Marguerite and the two small boys lived with them for the short time required for the proceedings. Walker wanted to have the boys spend the summers with him, but Marguerite, concerned that the woman he planned to marry would not be interested in or good to them, refused to agree. Walker was disappointed but accepted her decision, and Marguerite in turn was gratified that he did not precipitate a fight that would have been difficult for the boys. In the spring of 1934, Marguerite and the boys moved to Roanoke, Virginia, to be near Bill and Mamie Lee Andrews. They lived in the Andrews home until Marguerite could find a place of her own nearby. Neither the Sheltons nor the Andrews "took sides"; the divorce was "just one of those things." Both families continued their warm friendships with both Ken and Marguerite. "I am grateful for your openheartedness," Ken wrote Mamie Lee in December 1934. "I hope that you will accept my deep and sincere appreciation for the kindness and generosity which you have shown my boys and Marguerite."[12]

Kenneth Walker remarried; a son from the union was named John. John never got to know his father. His parents divorced shortly after his birth.

The last half of the thirties was a time when the Air Corps grew against a backdrop of increasing international unease. The General Headquarters Air Force set up in 1935 provided a reasonable framework within which to develop organizational and tactical skills; privately, some of the airmen agreed they were not yet ready for the full independence they sought.

At the ACTS, the doctrine of strategic aerial warfare continued to be refined even though it was not highly visible. Appropriations for personnel and materiel gradually increased as Hitler began threatening the peace in Europe. In 1937, the first B-17s, the plane that would make strategic air warfare possible, went to Langley and the 2d Bomb Group, commanded by Robert Olds. If the Air Corps is viewed as a three-legged stool, the legs being doctrine, control, and the means of implementation, the stool was sitting almost steady and level as the decade drew to an end.

After leaving Command and General Staff School, Walker went to Hamilton Field, California, where he served primarily with the 7th Bomb Group, commanded by George Stratemeyer. For a time, he was engineering officer for the

Walker (right) and Lt Col C. L. Tinker in 1931

11th Bombardment Squadron, which he had commanded at Langley. In October 1935, he became intelligence and operations officer for the 7th Group, a post he retained after becoming commanding officer of the 9th Bombardment Squadron in October 1936. In February 1938, he began three years' duty with the Hawaiian Department. Apart from a listing of specific assignments, the paper trail for those years is thin, the glimpses of Walker primarily anecdotal. Laurence Kuter, when updating Walker's bombardment text preparatory to teaching bombardment at the ACTS for the 1935 year, went to Hamilton Field to discuss the revisions with Walker, whom he knew to be active in new developments. The 7th Group at that time was equipped with a light bomber—the B-17s were not yet available in large numbers—but Walker advised Kuter that their light bomber experience had yielded nothing suitable for inclusion in a forward-looking text. Kuter nevertheless found Walker's review and comments helpful.[13]

Shortly after arriving at Hamilton Field, Walker cracked up a B-12A in landing at March Field, then under the command of Brig Gen Henry "Hap" Arnold. Walker's statement on the accident report was a terse, "overshot the flying field and hit rock with right wheel, breaking up right landing strut." Observers concurred that he came in too high and too fast, apparently because he had trouble getting his wing man in position. The accident was considered an error in judgment—his passenger could see no reason to prevent him from "giving it the gun." By this stage of his career, Walker had 3,360 flight hours. His station commander put in the record that since Walker had spent the four days prior to the accident doing "an unusual amount of work" with a wing maneuver at Salt Lake City, "excessive fatigue" possibly affected the situation. A fellow officer later remembered a 7th Bomb Group gathering when a skit roasted Walker for the incident. "They cited Ken Walker for extraordinary service . . . gave him some kind of a fur-lined thunder mug or something like that."[14]

For those trying to ready GHQ for a War Department service test, however, accidents were no joking matter. "As far as I can see," an exasperated Arnold wrote GHQ, "the only way to stop accidents is to keep all airplanes on the ground." He bewailed a recent rash of them. "Ken Walker," he wrote, "supposed to be one of our best pilots, apparently cuts out completely, uses up 4,000 feet and finally hits a concrete block and spoils a perfectly good airplane when he normally would have given her the gun and gone around again."[15]

Walker had another accident on 23 December 1937 while piloting a B-17, one of a hundred newly acquired. This accident happened on takeoff from the Municipal Airport in Denver, where plane and crew had stopped en route from Chanute Field to Hamilton Field. The local paper gave the crash a tremendous headline: FLYING FORTRESS CRACKS UP IN DENVER BUT CREW OF NINE ESCAPES INJURY. Takeoff appeared to be normal until, near the end of the field, Walker realized the plane was not going to lift into the air. He and his copilot, Lt William C. Capp, cut the engines, pulled in the retractable landing gear, and pulled back on the sticks to keep the tail on the ground and prevent a nosedive. After jumping a three-foot wire fence, the plane returned to the ground, skidded over a six-foot embankment and on for a half mile before coming to a stop in the middle of a highway. Walker's "presence of mind" and "expert maneuvering" were credited with preventing a major disaster. No one was hurt, and the major problem was figuring out how to get twelve and a half tons of airplane back to the factory for repair.[16]

In Hawaii, where GHQ Air Force was building up a composite wing, Walker's initial assignments were with bombardment. We see him as the operations officer for the 5th Bomb Group, signing a 40-page training directive that included a challenging tactical objective: "development and crystallization of the tactics and techniques necessary to insure the effective reconnaissance of sea areas, interception and destruction of a hostile fleet or elements thereof." The group's duties to that point had included such things as bombing a lava flow to protect a city, planting fig trees from the air, and rescuing men from the ocean. The training must have been sound; the onset of war in Europe found the group's personnel "excellently trained for warfare, but sadly lacking the modern tools of aerial war."[17]

An incident related by Gerald Robinson, then a new second lieutenant, illustrates that as group executive officer, Walker had relaxed and comfortable relationships with his men. An Army Day celebration was planned. Robinson and a buddy "got the brilliant idea that it would be great fun to make a parachute jump as part of the show." Neither had any experience jumping; nor did they have boots or helmet. Nevertheless, Walker approved their plan. Wearing the standard issue 24-foot emergency parachute, the men jumped from a B-18 "and entertained the guests." In the process, Robinson fractured a vertebra. "Ken Walker visited me in the Tripler Hospital often," Robinson wrote. "He was a very

B-18s over Mauna Kea and Mauna Loa in 1939

compassionate man, and never forgave himself for approving the jump. He used all of his influence to keep me on flying status."[18]

During Walker's years in the Hawaiian Department, the war in Europe became full blown. Hitler overran Poland in September 1939; after a winter of "Phony War," he fell on Europe in the spring of 1940. France capitulated on 25 June, and the Battle of Britain began in August. During these dramatic months, we see a frustrated Walker commanding the 18th Pursuit Group at Wheeler Field. Announcing the assignment, wing commander Brig Gen Walter H. Frank called Walker a "war veteran and aviation pioneer." He did not explain why a bombardment expert would now be given command of a pursuit group; he did say that Walker had been selected to attend the War College that year but had obtained permission to wait until 1941. Tremendous base expansion was underway in Hawaii, and we can speculate that personnel were in short supply. It is also possible that Walker wanted to obtain broader experience, as he had suggested would be wise in his 1933 letter to Spaatz. To Spaatz, he had deplored the "fetish of overspecialization," but the little evidence we have indicates that Walker did not enjoy his tenure in pursuit, which began in April 1940.[19]

Walker's adjutant at the 18th Pursuit Group was 1st Lt Bruce K. Holloway, who remembers the interlude well. "Ken had been in office for about a week, during which time he was in a dour mood; rather uncommunicative and certainly one not to be crossed—even unintentionally." Perceiving that Walker's assignment had been "a bitter blow," Holloway did his work and confined his remarks to "Yes sir" and "No sir."

"After a few days he said suddenly to me, 'If I am going to command this pursuit outfit I had better check out in the airplane, so come on. You can do the honors.'" Holloway eventually concluded Walker had none of the "emotional exhilaration toward flying a high performance machine that is so typical of fighter pilots," and this first episode was a case in point. The group had fairly new P-36s and the group commander's plane was a beauty, "manicured with multi-colored bands around the fuselage representing the three squadrons and a big number 1 painted on each side." It was known as the "Gold Bug." Walker was unimpressed.

"He was obviously impatient to get it over with," Holloway wrote, "and after I had tried to explain how to work the shotgun starter, he waved me off the wing and punched the starter button. The engine did not start, and before I could caution him to vent the gun breech before opening it, he opened it. The spent cartridge blew out from the residual pressure and ripped a hole in his trouser leg. This did not help matters at all, but after a brief [interlude] we inserted another cartridge and, this time, the engine started. He waved me off again and roared off into the wild blue yonder."

After 20 minutes or so, when Holloway was beginning to get nervous, he "heard a horrendous noise off to the south as though a whole flight of aircraft were in a steep dive." Holloway knew the sound—an unmistakable staccato caused by the propeller blade tips exceeding the speed of sound. For older aircraft with fixed-pitch blades, this could only happen

The "Gold Bug"

when the plane had reached very high diving speeds. With the new P-36, however, it could occur if the automatic propeller pitch control failed. The blade pitch would flatten out, relieving the load on the engine and allowing it to overspeed. The result would be a runaway engine and a frightful noise, plus greatly reduced forward thrust. It was an exceedingly dangerous condition. With sinking heart, Holloway realized that Walker was in grave trouble.

"Finally I spotted him. He was headed toward the field, but going painfully slow. I could feel a lump in my throat and said a silent prayer that the plane would not stall out. Somehow it did not, and to his everlasting credit he managed to get the wheels down and the airplane smoothly onto the grass runway in one piece."

Walker was visibly shaken. "Something really went wrong with this engine," he said. "It would only turn up 1200 rpm." Simultaneously relieved and upset, Holloway blurted out, "Did you notice whether the tachometer went all the way around before it stopped at 1200?" Afterward he realized his question was rude, but Walker seemed not to have heard. He never developed a fondness for his beautiful P-36. Some weeks later, one of his young flight leaders, David W. Hasher, lost his plane when another pilot taxied it into a ditch. Standing forlorn by the wreckage, Hasher felt a hand on his shoulder. "It was Major Walker. He was kind of smiling and said that as a flight leader, I needed an airplane of my own more than he did." They walked to the hangar together and Walker ordered Hasher's name put on the "Gold Bug." He may have been glad to see it go.[20]

Walker was with the 18th Pursuit Group only a short time, but during that interval Holloway "developed a high measure of respect and affection" for him, learning from him much that was invaluable in later assignments. "The ice was first broken," Holloway remembers, "when one day he said, 'Why did you ever want to be a pursuit pilot in the first place?' " Thereafter they often discussed airpower issues, and Holloway was impressed that Walker treated him as an intellectual equal. "This caused me not only to pay attention, but to offer counter comments. It was a truly cherished relationship."[21]

During his tenure with the 18th Group, Walker helped compile a group songbook, a collection of Air Corps songs old and new, sentimental and rowdy. "Air Corps tradition," he wrote in the introduction, "is predicated upon meeting grave and difficult situations with light hearts and high spirit; upon viewing possibilities of sudden death with detachment and levity. It is a tradition which carries us through trial and tribulation to ultimate successful accomplishment." One of the songs that Walker contributed was "The Student's Song," written by Walker and Hansell when they were together at the ACTS, with its plea, "We've got other courses to take-o/Just let up on us for God's sake-o." Another was "The Instructor's Lament," with a repeating chorus: "I don't want any more flying; I want to stay on the ground." Walker even managed to slip "The Bomber's Song" into the pursuiters' songbook. "Drown your sorrows and forget tomorrow's," it advised, "and drink a barrel to the old bombardment group."[22]

Walker's son Ken visited his father in Hawaii during that summer of 1940. Ken had last seen his father standing on the siding looking unhappy when he and his mother and brother boarded the train to leave when the family broke up. If Walker now hoped to warm up their awkward relationship, he was to be disappointed. Ken later remembered a few good spear-fishing trips together, but there were also promises of a trip to the island of Hawaii which never came off. Ken spent quite a bit of time swimming at Waikiki Beach, near his father's apartment, and Walker spent much of his time with a woman friend. Ken did not dislike her, but he did resent her since he himself received so little of his father's attention. A picnic for the three of them turned out to be tense; Ken had come back from a week at Boy Scout Camp without a merit badge, and his father expressed disapproval. After that, the visit went steadily downhill.

Walker shared little of his professional life with his son, although he did take him to see Wheeler Field. He also talked about one aspect of his Hawaii duty—daily reconnaissance flights over the Pacific, looking for incoming Japanese planes or ships. Less than a year later, after Walker had been transferred to Washington, the Hawaiian Air Force drafted a "Plan for the Employment of Bombardment Aviation in the Defense of Oahu." To provide reconnaissance, the plan called for a daily long-range search by B-17s with an attack force to hit anything found. The bombardment mission was "To attack and destroy enemy surface craft within radius of action." It recommended 180 B-17s and 36 long-range torpedo bombers for this task, on the grounds that it was, to their knowledge, "the best and only means that can be devised to locate enemy carriers and make attacks thereon before said carriers can come within launching distance of Oahu."[23]

Walker may not have worked on that plan directly, but it definitely reflects his thinking. He said much the same thing to his son that summer of 1940 and, after joining the Air War Plans Division in early 1941, he worked on plans for an Air Warning Service for Oahu.[24] Later, when the Japanese succeeded in their surprise attack on Pearl Harbor, Ken wondered why the reconnaissance flights his father had talked about had failed to spot the approaching carriers. (The reconnaissance plan was not implemented because it would have required more B-17s than the air forces had available.)

At the end of their month-long visit, father and son parted "rather formally." The groundwork for a close and meaningful relationship had never been laid, and Ken felt that he had visited a near stranger. Years later, the adult Kenneth Walker Jr. expressed certainty that his father loved him and his brother. Unfortunately, he seemed to have few ways of showing it—as though, not having had a close and loving father of his own, he lacked the basic understanding of how to be a good father himself. "He was a good father" Marguerite said in later years, "but his career came first." The primary way in which he related to his sons—expressing disapproval when they failed to meet his standards of performance—scarcely promoted warmth.[25]

According to Holloway's recollection of events, Walker was reassigned from the 18th Pursuit Group to another job in the islands late in 1940, but the two met again on 31 December. Having been reassigned to the mainland, they were among others leaving Honolulu on the US Army Transport *Republic*. Holloway and a friend, Charlie Robbins, decided the occasion required some special celebration. They organized a small galley brigade, armed with pots and ladles in lieu of drums, that marked the stroke of midnight by parading along the aisles of the stateroom decks and making a marvelous din. Holloway and Robbins were very shortly seized by two military police (MP) and put in the brig. The next morning, they were told to report to Major Walker.

"He gave us a first class chewing out," Holloway remembered, "and ordered us to report to him every two hours to account for our activities." Walker had gotten them sprung from the brig by assuring the transport's captain that the troublemakers would be strictly under his direct supervision until they reached San Francisco. He made certain the younger officers understood this, warning that "we would probably get keel-hauled if we screwed up again." The remainder of the voyage was uneventful. When Walker bade his charges good-bye in San Francisco, he told them—eyes dancing—"I suppose that maybe we have to put up with a few pursuit pilots after all."

It was the last time Holloway saw him, but he retained a strong memory of an officer whom he judged as "dynamic and indefatigueable, wiry, tough, smart, completely wedded to his convictions, hard-driving and of little patience for those who did not measure up to his standards of performance or application." Furthermore, Holloway testified, this dedicated bomber pilot even had "a good sense of humor."[26]

Notes

1. Kenneth N. Walker, "Bombardment Aviation: Bulwark of National Defense," *US Air Services*, August 1933, 15–19.

2. Donald Wilson, oral history interviews with Hugh N. Ahmann, K239.0512-878 and K239.0512-810, US Air Force Historical Research Center (USAFHRC), Maxwell AFB, Ala.: 43; Gen Laurence S. Kuter, oral history interview with Hugh N. Ahmann and Thomas A. Sturm, 30 September–3 October 1974, K239.0512-810, 129–30.

3. Maj Gen Howard C. Davidson, oral history interview with Hugh N. Ahmann and Thomas A. Sturm, 5–8 December 1974, K239.0512-810, USAFHRC, 385.

4. See Walter Millis, *Arms and Men: A Study in American Military History* (New York: Putnam's, 1956), 250–59, 269-70; and Michael S. Sherry, *The Rise of American Air Power: The Creation of Armageddon* (New Haven: Yale University Press, 1987), 50.

5. Bob Pearson to Douglas Walker, letter, 1 September 1964.

6. Federal Aviation Commission Testimony, 248.121-3, USAFHRC, and Report, 248.211-53, USAFHRC, 1, 4, 120, 253-54. See also Robert Frank Futrell, *Ideas, Concepts, Doctrine: Basic Thinking in the United States Air Force, 1907–1960* (1971 reprint, Maxwell AFB, Ala.: Air University Press, 1989), 37–38.

7. Bill Andrews to Douglas Walker, letter, date unknown. Other material from Mamie Lee Andrews, telephone interview with author, 28 August 1990.

8. Hansell, Haywood S. Jr., "Brig. Gen. Kenneth N. Walker, Prophet of Strategic Airpower," *Air Force Magazine*, November 1978, 92–94.

9. Douglas C. Shelton to Douglas P. Walker, letter, 18 January 1991.

10. Claire Lee Chennault, *Way of A Fighter* (New York: G. P. Putnam's Sons, 1949), 18.

11. Mamie Lee Andrews, telephone interview with author, 28 August 1990; K.N. Walker Jr. to author, letter, 8 March 1991; and Author conversations with Douglas Walker.

12. Marguerite Walker, videotape interview; Shelton to Walker; Mamie Lee Andrews telephone interview with author, 28 August 1990; Ken Walker to Mamie Lee and Bill Andrews, letter, 20 December 1934.

13. Laurence S. Kuter, unpublished autobiography, US Air Force Academy Library, Special Collections, 122, 125.

14. Accident Report, 200.3912-1, USAFHRC. Davidson, interview, 385.

15. DeWitt S. Copp, *A Few Great Captains: The Men and Events That Shaped the Development of U.S. Air Power* (Garden City, N.Y.: Doubleday, 1980), 308–9.

16. *The Denver Post*, 23 December 1937, 1, 3.

17. Training Directive, 1 July 1938–30 June 1939. GP-5-SU-DI(BOMB), USAFHRC. GP-5-HI(BOMB)1918-45, USAFHRC.

18. Gerald G. Robinson to author, letter, 17 January 1991.

19. GP-18-HI, newspaper clippings 22 March 1940, USAFHRC. Copp, 260.

20. David W. Hassemer, San Antonio, Texas, to author, letter, 21 January 1991.

21. Bruce K. Holloway to author, letter, 19 July 1990, including unpublished manuscript, "Wheeler Field, T.H.: 1940."

22. Songbook in Walker's Medal of Honor Scrapbook, Office of Air Force History (OAFH).

23. AWPD, Plan for the Employment of Bombardment Aviation in Defense of Oahu, 14 August 1941, 145.811-25, USAFHRC; and Wesley Frank Craven and James Lea Cate, eds., *The Army Air Forces in World War II*, vol. 1, *Plans and Early Operations January 1939 to August 1942* (Chicago: University of Chicago Press, 1948): 174.

24. Memorandum, Maj K. N. Walker to Gen Carl Spaatz, War Plans Division, subject: Committee to Discuss Air Warning Service, 21 February 1941, AAG cl. file 676.9 A, Aircraft Warning Service Site, partial copy in Walker's Medal of Honor Scrapbook, OAFH.

25. K.N. Walker Jr., to author, letter, 8 March 1991. Marguerite Walker, videotape interview.

26. Holloway to author, letter, 19 July 1990, including unpublished manuscript, "Wheeler Field, T.H.: 1940."

Chapter 4

Washington and AWPD-1

Walker went to Washington in January 1941 to take up duties in the Office of the Chief of Air Corps as Assistant Chief, Plans. It must have seemed somewhat like a homecoming, for Brig Gen Carl A. "Tooey" Spaatz was head of Plans and two of his assistants were Robert Olds and Muir Fairchild, old friends from the ACTS. The Air Corps offices were in the old munitions building on Constitution Avenue, along with other War Department personnel. Before long, Walker was a regular figure at a cafeteria called the Allies Inn, where the airmen had a table for lunch that attracted those who shared their professional convictions. Newcomers were sometimes startled by the intellectual quality of the conversation.[1] One thing the Air Corps Plans Division was working on when Walker joined them was a reorganization that would give airmen some independence in deciding how to run their branch of the Army; luncheon conversations were doubtless lively.

Walker worked on one topic that was highly secret and probably not discussed at the luncheon table. He provided guidance and help to Richard Aldworth, a retired captain serving as vice president of a little-known company named Central Aircraft Manufacturing Company (CAMCO). CAMCO was working with Chennault, Chinese officials, and a few Americans on a highly secret project that had the President's blessing. The goal was to send American planes and pilots to China to take part in their war against Japan. Part of the plan matured into the American Volunteer Group, a fighter group later famous as Chennault's Flying Tigers. Largely unknown at the time, the total plan also included a bomber group. In compiling personnel requirements for such a group, Aldworth consulted the Air Staff. "I suggest before you make [a] decision on the bombardment personnel," Aldworth wrote presidential assistant Lauchlin Currie, "that you consult with Lt. Col. Kenneth N. Walker." Aldworth attached to his overall study a letter from Walker, whom he described as "one of the foremost authorities on bombardment in the United States." Unfortunately, Walker's letter was not found in the file.[2]

By the time Walker joined Plans, war was raging in Europe and in China. Most Americans seemed more determined than ever to stay out of it, but President Roosevelt and a number of others realized that US vital interests were, indeed, at stake. Consequently, a posture of aggressive defense had begun taking shape after the Munich conference in September 1938 (when the Allies acquiesced to Hitler's conquest of Czechoslovakia). Munich gave the Air Corps a new status, for Hitler had won his victory largely through the threat of airpower. Ambassador William C. Bullitt told President Roosevelt pointedly, "If you have enough airplanes you don't have to go to Berchtesgaden."[3]

Roosevelt got the point. By October 1938, he was talking about expanded air forces. In November, he said he wanted 10,000 planes and plant capacity to build 10,000 per year. In January 1939 he told Congress, which had to appropriate funds, "survival cannot be guaranteed by arming after the attack begins, for there is new range and speed to offense." The money was forthcoming. The 1939 expansion approved a total Air Corps of 5,500 planes (up from 1,700). The officer strength was doubled (to 3,202) and enlisted men went up 150 percent (to 45,000), while a record $7.5 million was allotted for research and development.[4]

At the head of this suddenly respectable branch of the Army was the smiling powerhouse, Brig Gen Henry H. "Hap" Arnold. Several changes in organization took place in 1939, but they failed to solve the basic problem of command. In March 1941—about the time the Lend Lease Act was going into effect—Secretary of War Henry L. Stimson ordered that air be put under a single commander. Guided by Robert A. Lovett, Assistant Secretary of War for Air, the US Army Air Forces (AAF) took shape—not the complete independence airmen had sought, but a substantial step. Both the Air Corps and the Air Force Combat Command (replacing the GHQ Air Force) were placed under the new AAF. Arnold, as its head, had responsibility for policies and plans for all Army aviation and was directly responsible to the Army's chief of staff—by this time General George C. Marshall, who perceived a valuable potential for airpower. Arnold continued to be deputy chief of staff, thus connecting the AAF to the War Department General Staff. During coming months, even this step toward independence was to prove insufficient; greater autonomy would come in March 1942. Under this interim organization (which took effect in June 1941), however, airmen produced the basic strategic plan that guided US air forces throughout the Second World War.[5]

The 1941 reorganization allowed Arnold an air staff comparable to the Army's general staff. Arnold immediately began shuffling personnel, naming Spaatz his chief of staff and transferring three lieutenant colonels to his new Air War Plans Division (AWPD). One of them was Walker, who became the sole member of the division's War Plans Group. As chief of the new AWPD, Arnold called on Lt Col Harold George, then at Langley commanding the 2d Bombardment Group (the only air unit equipped with B-17s). George reported to Spaatz on 10 July.

The day before, 9 July, Roosevelt sent an important letter to his Secretaries of War and Navy, Henry Stimson and Frank Knox. He asked them to explore "at once the overall production requirements required to defeat our potential enemies." From their report, "we should be able to establish a munitions objective indicating the industrial capacity which this nation will require."[6] The request went to the Army's War Plans Division through Stimson. Meeting it promised to be a tremendous task—for before they could determine what weapons and munitions should be produced, the planners must first define the strategic concept of how potential enemies would be defeated and what military units, with what sort of equipment, would be needed. They had been working on the issue for some time, but time was running out: The president wanted an answer by 10 September!

The international outlook was grim. The Germans had taken over the Balkans and were threatening to do the same in North Africa. Late in June, they invaded Russia and most observers feared that she would quickly succumb. Japan threatened further aggression in the Far East, raising the strong possibility of a two-ocean war. The foundation for planning such a war had been laid, however, in Rainbow-5, already approved by the Joint Board. In addition, joint staff talks with the British during the early months of 1941 had produced ABC-1, a significant agreement on basic Allied cooperation and strategy should the United States enter the war. According to ABC-1, Germany was the primary enemy; if war with both Germany and Japan ensued, a strategic defense would be maintained against Japan until Germany was defeated.[7]

Within this framework, the Army's War Plans Division (WPD) set to work on the President's request. Maj Albert C. Wedemeyer was placed in charge. Well-read and intellectually oriented, he realized that the Army as it then existed could not meet the crisis. It had to be rebuilt as a mechanized force, fully mobile, equipped for antitank and antiaircraft defense, and with the armored division as the main offensive tool. Using historical experience as a guide, Wedemeyer estimated that the United States could mobilize approximately 14 million men without disrupting the industrial base which would be necessary to sustain them in combat. But the available Axis manpower was far greater. It was obvious that the gap must be closed by equipment, including the airplane.[8]

One basic planning approach was to analyze the forces of the potential enemy and by comparison arrive at US requirements for surface forces necessary to defeat them. The air forces, however, offered no feasible point of comparison. Bombers are not pitted against bombers, nor had there been enough aerial warfare in the past to provide guidelines about how much force was necessary to accomplish a specific mission. Those working on the air annex for the overall plan made slow progress.

Late in July, Lt Col Clayton L. Bissell, an air officer in WPD, suggested to George that the new AWPD assist the air officers of WPD in drawing up the air portion of the study. George knew opportunity when he saw it. As recently as 1938, the Air Corps had been denied funds for developing a long-range bomber and told to confine their research to craft for close support of ground forces. Fearing that if the Army now controlled this fundamental production plan, it would not include planes for carrying out the strategic mission airmen believed should be pursued, he objected to Bissell's proposition. AWPD should *prepare* the air annex, he said, rather than assist WPD. He persuaded Spaatz, who persuaded Arnold, who took it up with the head of WPD. "Go ahead," he was told. So, on 29 July, Bissell turned over to George the complete file with the reminder that they must conform to Rainbow-5 and ABC-l. Those were the only restrictions.[9]

At this point, AWPD was less than a month old and had only three officers for the planning task: George, Walker, and Maj Haywood Hansell. The three were united in friendship as well as profession. Each had participated actively in air's doctrinal evolution during the preceding decade; each had taught at the ACTS; each held the vision of the strategic air mission based on bombardment. Their particular strengths complemented each other. George's political skills and smooth manner were enhanced by his legal training (he held a law degree from George Washington University); his recent experience with the B-17 gave him first-hand knowledge of Air Corps capability. Walker was the backbone of bombardment, the believer, the aggressive advocate who brooked no doubts and drove policy as relentlessly as he drove himself. Personal relationships among the men were good. When Walker made out his will a few months later, he named George the executor of his estate. Hansell looked up to Walker—"adored him," according to some[10]—and brought to the team a lively zest, a moderating humor, and a shrewd mind. His most recent assignment had been in air intelligence; he was just back from England with a plane load of valuable information on target data from the Royal Air Force (RAF).

When these three sat down to analyze the task ahead, Walker suggested they request the temporary assignment of Laurence Kuter, then a major in the operations division of the War Department General Staff. This was done. He too had instructed bombardment at the ACTS; he was philosophically at one with the others. He strengthened the team with an intellect described as cold, efficient, balanced, and persistent. Less than a year later, he would become the Army's youngest general at 36. An acetylene torch, one writer called him: hot enough to cut steel but never burning out of control.[11] The future of US airpower now rested on these four men, hard working visionaries who had devoted their entire professional life toward this moment. "We realized instinctively that a major milestone had been reached," Hansell wrote. "Suddenly, without anywhere near the opposition we expected, we found ourselves able to plan our own future."[12]

They had little time. Arnold left Washington on Sunday, 3 August, to take part in the historic meeting of Roosevelt and Churchill at Argentia Bay. He left word for George to have the annex ready by 12 August. This gave them only nine days.[13]

By this time, however, George had already figured out what they would do and how they would divide the work. He would direct it himself. The War Plans Division had asked them to determine the maximum number of air squadrons the Army Air Forces might require, but their report would go further. It would include strategy, timing, and targeting, as well as production, manpower, training, organization, support, and basing for air forces both in the United States and abroad. This was divided into 18 specific topics, each of which would make a separate tab to the completed annex. At least two officers would work on each tab. They could get some temporary help from air officers in other departments, but the major responsibility would fall on the four of them.[14]

More agonizing was the next decision: how should they state their basic strategic concept, and how far could they go without antagonizing the War Department and thus losing everything? They had a starting point, for among the offensive policies agreed upon in ABC-1 was a sustained air offensive by US and British air forces against Axis military power at its source.[15] George discussed the issues with the others, then made the decision: they would outline a strategic air offensive against the German war machine and economy. Such an offensive would be essential preparation for an invasion of the continent, but their plan would not rule out the possibility that the air offensive alone might bring Germany's defeat. There were no precedents to help them, no guidelines from experience for determining methods, objectives, or targets. They had only their own resources, refined and firmed up by their theoretical analysis at the ACTS and their own practical experience.[16]

By Wednesday, 6 August, George had completed the summary of their basic decisions. The four men met with the other officers who would assist them. Walker and Kuter went through each tab of the plan individually, explaining what should be included. The air effort was divided into five planning tasks, each composed of a number of parts. The first task was to conduct an air offensive against Germany and Italy to destroy their will and capability to continue the war and to make an invasion either unnecessary or feasible. Having noted the stiffening resistance of the British under German air attack, they discarded the concept of destroying the enemy's will to resist by bombing cities. They would rely on the precision bombing of systems that were vital to Germany's ability to fight.[17]

The second and third tasks were to provide air operations in defense of the Western Hemisphere and a strategic defense of the Pacific. The fourth was to provide close air support of the surface forces in the invasion of Europe and the major land campaign thereafter. The fifth task was to calculate the total air requirements to accomplish all the preceding. The final plan would include all their supporting calculations so there could be no doubt as to how conclusions had been reached.[18]

Each of the four officers had primary responsibility for several specific topics, although there was a great deal of interplay among the entire group. Walker was in charge of "Bombardment Operations Against Germany," "Aircraft Required for Control of the Seas," and "Bombardment Aviation Required for a Strategic Defensive in Asia." Kuter, who respected Walker a great deal, was his associate on all three. Walker was associate with Kuter for "Bombardment of Operating Bases" Walker was also responsible for "Escort Fighters Needed to Support Bombardment Operations" (with Lieutenant Colonel Schneider and Major Vandenberg). Any one of the above topics would have been a nine-day task. The working outline for "Bombardment Operations Against Germany," for instance, listed these subtopics: the means and methods of operating against the German Air Force; calculations of force required to destroy all vital points on Germany's inland waterways, with similar calculations for destroying gasoline and oil production, rail communications, and power plants; a discussion of other vital objectives; estimated effects of rendering the vital establishments inoperative by 50 percent; total bombardment planes required; and estimation of attrition rate and replacement aircraft required. Even to arrive at this outline required the critical basic decisions as to what main targets would have the most potential for crippling Germany's ability to fight. The years of study and analysis and theoretical projection helped the team focus on power, transportation, and oil. Within those areas, they pinpointed 154 selected targets which they believed would neutralize Germany's war-making capability if they could be destroyed and kept out of operation. Their schedule allowed a year for production, training, and organization, and an additional nine months for deployment, which would be followed by a full-scale aerial offensive of six months.[19]

The next step was to determine how much force—how many planes and how many bombs of what size—would be required to accomplish a degree of destruction that would meet the objective. Of necessity, much of this was based on mathematical calculations and probability because wartime experience with US planes was unavailable. But Walker had worked on this problem as early as 1930, when he wrote his bombardment text; with his methodical intensity, he dug in. On 30 July, he tackled the overall problem, discussed it with others, and secured reports from the Air Corps Board pertaining to range, direction, area, and bases of calculation of probability of hitting. The next day, he requested from the Statistics Section some specific graphs that would take a number of variables into consideration and speed up their calculations. They had data from the RAF to help them compute bombing accuracy under combat conditions. Walker prepared an outline to be used as a basis for estimating the force required.[20]

Step by step, the material came together. The pressure was intense, the weather hot and humid, the working conditions abysmal. The Pentagon was then in the planning stages; the 1941 staffs worked in the crowded and inefficient munitions building. AWPD was housed in its latest and uppermost addition, the Penthouse, where heat was at its worst. The first weeks of August were incredibly hot; in lieu of air-conditioning, fans moved the hot air around. "When you put your hand down on your desk," Hansell remembered, "your papers would stick to it."[21] They worked early and late in this atmosphere, with only essential time off for meals and sleeping. George, who kept a controlling hand over the entire process, sent a reminder note on Friday, 8 August, that the final copy must go to Arnold by Tuesday; those responsible for the tabs should meet with him in the War Room at 1330 on Sunday with their final computation on the required numbers of combat aircraft.

The figures were staggering: 61,799 combat aircraft, with total personnel at 2,164,916. More planes and men must be supplied as replacements for losses; the attrition in aircraft was estimated at 2,133 per month.[22] Considering that, less than two years ago, there had been opposition to increasing the Air Corps to 5,500 planes, dare they proceed with their current plan? Even as they were developing it, Congress was locked in a heated debate on whether or not to extend the draft (it passed on 12 August by only one vote). Were the AWPD planners being realistic? Kuter later said that they laid their plans as though the planes were already at hand. But was it even possible that such numbers could be built? If ever the courage of conviction was critical, this was the time. They acknowledged the problems, assumed that the numbers were attainable, then completed the plan. "We had to win a war," Kuter said. "There was no lack of confidence."[23]

Throughout his career, Walker had been confident, the one most firmly believing that the bombers would get through. Now, one of his responsibilities was the tab, "Escort Fighters Needed to Support Bombardment Operations." The worksheet for discussion listed four issues.

> 1. Necessity with respect to insuring daylight operations.
>
> 2. Ranges. Types of aircraft. Proportion of escort fighters to bombers.
>
> 3. Determination of escort fighters required if their employment is deemed essential to insuring effective daylight operations.
>
> 4. Tabulation of numbers required by July 1, 1943. Based on attrition rates, the monthly replacement rate from the later date on.[24]

Unfortunately, there was little else on which to reconstruct the group's thinking; and of all the momentous decisions they made, those concerning escort fighters would later prove most controversial. During that August of 1941, did Walker recall Chennault's dogged insistence that defensive fighters would make scrap metal out of attacking bombers? Had he read and pondered intelligence reports from Spain in 1936–37, or China in 1937–38, about bombers and fighters in action? The conclusion of those involved in the Sino-Japanese War was that escort fighters were essential. Some American analysts considered those experiences inapplicable to US circumstances; but 1937 lectures at the ACTS, based on the Spanish Civil War, concluded that bombardment operations facing hostile fighters needed pursuit protection as well as heavy defensive armament and tight defensive formations. Pilots who had fought in Spain testified that the only effective protection for bombers was the single-engine pursuit; the "flying fortress," they said, existed only in the minds of theoreticians.[25]

By the time of the Spanish and Chinese battles, Walker had left the ACTS. He may not have spent much time studying those experiences, but certainly he and the other AWPD planners had followed the aerial warfare in Europe thus far. Its most dramatic aspect had been the German use of air in support of ground forces, an application of airpower the Air Corps had resisted. Attack aviation, as well as pursuit aviation, had ranked low in Air Corps priorities. After the German blitzkrieg in Poland, however, Arnold stated bluntly that the Air Corps position (that fighter aircraft could not shoot down large bombers) had been proven wholly untenable. He asked GHQ for a study; the response was that the present bomber could not defend itself adequately against pursuit. During 1939 and 1940, other airmen also began to rethink the basic concept of bomber invincibility—that rock on which Walker had stood since 1930. Some noted with concern that the RAF lost so heavily to the Luftwaffe in daylight bombing raids over the continent that they had to abandon them for less accurate night raids. Conversely, British pursuit inflicted substantial losses on German bombers and pursuits over England. The Air Corps Board recommended putting a rack on pursuit aircraft so they could carry either a bomb or a droppable fuel tank. It also recommended that consideration be given to development of a long-range fighter. In 1940, George told Arnold that it looked to him as though the bombers were going to need fighter protection.[26]

Basically, however, the ACTS position held firm. American bombers were better armed; American formations were tighter. It was still possible to believe that the bombers would get through, that they could get through in daylight to conduct precision raids, that enemy air strength could be destroyed on the ground by bombing installations and factories, that pursuit's role would be primarily defense of bases.

This was the position now taken by the AWPD planners. "Each of us," Kuter wrote some years later, "had scoffed at the idea that fighters would be needed to protect bombers, to enable bombers to reach their objective. In preparing AWPD-1, we stayed in that rut." He deemed it "harsh justice" that Walker was later killed in an unescorted B-17, while he and Hansell each had the agony of commanding under conditions of high loss before escort planes could be provided.[27]

There were some small nagging doubts. In a memo to Arnold on 11 August, Hansell admitted that while the B-17 provided the means of coping with fighters "for the moment," it would in the future need additional firepower.[28] And in AWPD-1, the planners went one small step farther. To guard against expected improvements in German fighter defenses, the plan recommended that experiments be begun immediately to develop a heavily armed and armored escort fighter with long-range capacity. The plan called for additional development and an eventual force of 3,740 bombers with a 10,000-mile range—a force that would be essential should Russia or Britain be defeated.

But the plan called for only 13 test model fighters to accompany them. It is hard to escape the conclusion that they expected little to come from the development efforts. Hansell, trying to explain why a later version of their plan (compiled after Walker had left) eliminated even this limited request for escort fighters, wrote that there were no escort fighters in existence and it was not considered possible to design and produce them in the short time necessary.[29] Bissell, who attended several of AWPD's planning sessions that August, argued that the need for the escort fighter was just as great as the need for the bomber—and just as feasible technologically. Recognizing a possible need, however, was apparently as far as Walker and the others were willing to bend.[30]

The influence for that small bend, which Hansell deemed significant because it was the airmen's first official recognition of the need for escort planes, may have come from George and from Maj Hoyt Vandenberg. George had already made his reservations known, while Vandenberg, Walker's associate on the escort planes tab, had supported the need for pursuit during the long arguments at the ACTS during the 1930s. Walker himself seems not to have wavered in his conviction that the bombers would get through. A friend who often saw him during these months described him as "a man who knew what he was doing, and wanted things done his way." Walker believed the problems the British were having could be solved by bigger and faster bombers that could fly higher and mount more guns. What was needed was to train gunners properly, tighten up bomber formations, and set your mind to the target. Pursuit remained at the bottom of his priority list. At one stage during their planning, when they tackled the problem of too many planes for the available air bases in Britain, his approach was to build as many new bases as possible and cut the number of fighters if something had to give.[31]

AWPD-1 was completed and turned in to WPD at midnight of 11 August. Hard-pressed to complete their own portion of the plan, WPD simply appended it to their own work as "ANNEX 2, Requirements of Army Air Forces." But even as the complete plan went to the Government Printing Office for reproduction, George realized they now faced their biggest challenge: selling their plan to the War Department. Although their plan included the tactical aircraft that the Army would expect and want, the strategic air mission which they had incorporated not only went counter to prevailing plans for the use of airpower but placed the Army's ground forces in a secondary role, slated to go in only after air had paved the way, if at all. The airmen had much to lose if they failed to get their plan adopted.

George's legal mind was up to the challenge. He arranged a first briefing with Kuter's commanding officer in the Operations Division; he was sympathetic to their ideas and had a good relationship with Kuter. Walker had suggested that people who would quibble about numbers were people who had already bought the plan, so George prepared an introduction full of statistics and figures, with the controversial priority assigned to the air offensive falling into place without fanfare. After his introduction, Hansell, Walker, and Kuter spoke in turn, presenting different portions of the plan. Their initial presentation went well. George then insisted on a carefully scripted refinement. He wanted the presentation to be polished—as professional as the work that had gone into it. At his insistence, each of them wrote, timed, and memorized his part verbatim, after which they rehearsed the complete product, speaking without notes and using maps and charts to illustrate their points. The entire presentation took two hours. Walker, whose unequivocal belief in what they were proposing had sustained them all throughout the nine days of intense preparation, memorized his part along with the rest. He believed in their plan; he would do his part to sell it.[32]

On 13 August, the planners began a series of presentations—eight of them within a month. Among the first to hear it were Arnold and Lovett, whose enthusiastic reception raised morale all around. On 30 August, Marshall heard it, sitting silent until the end, when he said he thought it had merit and should be presented to Stimson. Morale jumped again, for Marshall's support was critical. The briefing for Stimson was scheduled for 12 September. On the afternoon of the 11th, he called George, Kuter, and Walker to his office for an informal chat, possibly to ferret out any rough edges that their formal presentation might smooth over. For an hour and thirty-five minutes, the three answered specific questions about a number of details. They had no quarrel with Stimson's major observation that the enormous expansion they proposed could not take place unless the nation was at war or in a war spirit.

The following morning, Stimson and Assistant Secretary John J. McCloy interrupted a prearranged schedule and listened to three parts of the formal presentation, including Walker's on Bombardment Force versus Germany. Afterward, George assessed that Stimson "accepts the study as a matter-of-fact statement of the force required to do the job." As for McCloy, he was pleased at AWPD-l's "offensive nature."[33]

The airmen had pulled it off. A decade of doctrinal debate had been converted into a specific war plan and accepted by the War Department. Airpower was to have its test.

There was no letup in the work schedule that fall, for tensions continued to mount as the nation moved inexorably toward open warfare. Planning staffs worked feverishly to keep abreast of changing needs. It is difficult to follow the work of any individual in Washington's complex maze, but it is known that Walker contributed to AWPD-2, drawn up on 9 September to allocate airplane production from 1 October 1941 to 30 June 1942. It is possible, although unlikely, that he wrote the unsigned memo on airplane requirements that went to the Air Corps Materiel Division on 12 September; the memo stressed the importance of a bomber with an action radius of 4,000 miles, and emphasized: "It is mandatory that escort fighters be developed for test without delay. An escort fighter with a range comparable to the bomber it supports must be developed to insure day bombing missions in spite of opposition by the pursuit developments expected in the near future."[34]

One problem planners faced was too few planes coming from the factories to meet competing needs—building up the US forces, plus supplying Lend Lease aid to Britain, Russia, and China. On 25 September, Walker met with the President's trusted confidant and assistant, Harry Hopkins, to present the Air War Plans point of view. They noted that "eventual defeat of the Axis would probably require American intervention; that unless we could organize our combat units for operations, we would delay indefinitely the creation of the trained force essential; and that it appeared that if our point of view is at all correct, we should give the minimum aircraft to the nations opposing the Axis which would enable them to keep going and devoting the rest of it to building up an offensive Air Force." Walker noted that his argument "seemed to interest Mr Hopkins."[35]

Decisions on reinforcing Panama, Hawaii, and the Philippines highlighted the need for unity of command, but the traditional turf wars between Army and Navy proceeded nevertheless. We see Walker in late October contributing his bit, arguing against the Navy's revision of a plan and insisting,

"The Army must lose no opportunity to insist upon exercising its right to conduct air operations within the tactical operating radius of its aircraft—as an Army responsibility. It must be shown that the Army does and can operate in lieu of naval forces." He made it clear that the Navy must not "cut into our organization of our heavy bombardment groups" by substituting four-engine bombardment-type airplanes for patrol boats.[36]

Walker's impassioned advocacy of the air forces came up again in November; but on this issue, he obviously put defense foremost. When the Alaskan Defense Command sent Army bombers on offshore patrols, the Navy protested that this was a naval responsibility, even though the Navy had no planes in Alaska that could conduct such patrols. When Walker drafted AWPD's reply, this often abrasive officer summoned a surprising reservoir of tact and diplomacy. Offshore patrols were a Navy function, he said, but Army pilots were required to carry out overwater reconnaissance to keep up their combat proficiency. To avoid conflict, these should henceforth be called "tactical reconnaissance" instead of "offshore patrol."[37]

We do not know Walker's reaction when, on 4 December, the *Chicago Tribune* and the *Washington Times-Herald* published the plans on which War Department personnel had worked so hard. Stimson thought the scoop was absolutely evil; but Hitler, who could have profited from its information, refused to follow the advice his generals compiled from reading it.

After Pearl Harbor, when AWPD drafted AWPD-4 to fit rapidly changing circumstances, Walker again worked on the plan. Differing primarily in degree from AWPD-1, it called for an air force of 3,000,000 men and 90,000 planes, these to be produced by giving "national first priority to the production of aircraft." When British and American planners came together at the end of the year, by then Allies in fact as well as sentiment, they adopted AWPD-1 as the most realistic plan for the existing circumstances.[38]

On the evening of 7 December 1941, as Americans everywhere were reeling with shock from the news of the American defeat at Pearl Harbor, Walker put on his pinks and greens, tucked his swagger stick under his arm, and went to the home of his closest Washington friends to tell them that he would be leaving for the fight.

Kenneth Walker's personal life during the months he served in Washington was dominated by his love for Cleo, the woman he met while in Hawaii, and his friendship with C. L. and Mary Hodge. Walker met the Hodges at a social gathering early in 1941. Hodge was at that time the chief economist with the State Department while Mary ran their large and gracious home in Arlington, a home that was always open and welcomed many people. An immediate friendship developed, for in Walker the Hodges saw "a wonderful man, pleasant and knowledgeable, the sort of man you like to have around." Cleo came to Washington to be near Walker, and when Mary Hodge met her and learned she was looking for a place to stay, she invited her to live in their guest suite. Cleo accepted. During the coming months, Walker was in the Hodge home a great deal—often with Cleo, but also as a close friend of the Hodges.[39]

"He was like one of the family," Mary Hodge recalls, "and we spent many hours in a very homey atmosphere." Walker and C. L. Hodge had many things in common and held many long talks. Mary Hodge, who remembers that Walker told amusing yarns—"sometimes pretty raunchy"—and that conversation tended to be lively and about "almost everything." When the topic came around to war, Walker would get "wound up" and use language that was "rather strong." She saw him as "intense, high strung, all that sort of thing." "Some people like that are not very pleasant to be around," she admitted, "but you liked Ken. He was gregarious and social—pleasant to be around." She responded to his good looks and sharp dress; she enjoyed his sense of humor and his teasing. "It was a nice, easy, friendly relationship; he knew he was welcome here."[40]

And what of Cleo? The Hodges' daughter Mary Lee, 12 at the time, later recalled her as a "babe," a flashy dresser with gorgeous legs but "not the sort of woman who went to State Department dinners." Mary Hodge, however, describes her as "a very nice person," tall and with stunning good looks, younger than Ken, and very much in love with him. Mary

Hodge sensed that Walker's tendency to be autocratic meant he would have demanded a great deal of a wife; she speculated that he could have been "very difficult" to live with. At first, she did not know that he had been married before. One night, her husband and Ken stayed up quite late, talking about his earlier marriages and their failure; but her husband told her only that the conversation had been confidential and that he was "satisfied." Her own assessment was that Ken was not haunted by his past relationships, that he felt good about himself, that he was very much in love, and that a marriage between Ken and Cleo might have worked well. "She handled him well, and he didn't know he was being handled," she said.[41]

Walker did not introduce Cleo to Mamie Lee and Bill Andrews, who then lived in Bethesda. Nor did he take her to the home of Muir and Florence Fairchild, where friends sometimes gathered for social events. Mamie Lee understood, however, that he was seeing someone. She remembered one evening when he came to the Fairchilds' "dressed up, with a Chesterfield collar and a swagger stick," and her laughing reaction was that he was "going to the dogs, he was so dressed up." Ethel Kuter, who described the atmosphere in Washington during 1941 and early 1942 as "not intimacy," a time when social life "pretty much disappeared," does not remember Walker being in their home.[42]

On that Sunday of Pearl Harbor, as Walker stood before the large windows in the Hodges' living room and talked about the attack and the war ahead, he created for 12-year-old Mary Lee "my most vivid memory of the war." He was solemn, as befitted the drama of the moment; it was obvious that he wanted to go, and soon. As he talked, Mary Hodge was struck by how "very military" he was. She had realized that he was very wrapped up in his profession, although conversations about the military had usually been with her husband rather than with her. She had sensed that he would "probably be happier in a plane than on the ground," but now she began to realize how important his military life was to him.

During the coming months, Walker chaffed, for not until June 1942 did he receive orders to report for combat duty. In March, he wrote Ken Jr., expressing the fighter's weariness with staff duty and saying that the new War Department organization "doesn't please us very much." "I shall have to go up to the new superior general staff—in the War Plans Division. I'm pretty unhappy over it." He told his son he would "try very hard to remember your 'pep talk' about 'they also serve who only sit and wait' or something like that, and try and get some comfort out of the thought."[43] Walker may have been chaffing at his desk job, but he performed it well. After his death, he was awarded the Legion of Merit in recognition of his contributions as a staff officer. He not only contributed to AWPD-1, but to the reorganization of the air forces that was effected in March 1942.[44]

In another letter to Ken Jr. in May, Walker grumbled a little about "getting to be quite a Kiwi—for I don't get out among combat outfits and don't often see our later planes." At that time, he was "still expecting to get active service but am afraid I may have to wait for a number of months. A couple of generals are trying to get me pried loose from the General Staff but no luck so far."[45]

In both letters, Walker revealed his concern that his boys achieve. "Get the eagle rating," he urged, and "keep the old nose to the grindstone to the extent of getting the school work done." He noted that Ken's grades were "OK," but added, "good grades do not indicate necessarily what one learns." The real test, he insisted, "is the test one gives himself—'am I getting the most out of my opportunities.'" He urged them to learn to box and be good at it, but reminded them that he had gone in for boxing but couldn't keep it up, so he suggested they take up tennis as "something that one can do all his life." He told Mary Hodge the same thing he had told Mamie Lee Andrews—that he hoped his sons would have a better life than he had had. He did not elaborate, and she wondered what he meant because to her he seemed thoroughly at ease with who and what he was, self-confident with what he had achieved in his career. Perhaps the demons that drove him could not be satisfied.

In late May or early June, while Walker waited for his orders, his old friend Bob Pearson came to Washington on a business trip. He brought with him his son, Scott, newly graduated from high school. As a special celebration for the boy, the Pearsons met Walker at the Willard Hotel for a farewell lunch. Scott thought Walker cut a great figure that day, complete with riding crop and leather boots, but the tenor of the conversation was "resigned," and Walker's "state of mind regarding his personal life was quite negative." Standing on the sidewalk outside the Willard after lunch, Scott listened as Walker told his friend, "I've made a terrible mess of things over here. I doubt if I'll be back." Fifty years later, Scott found the drama of the scene still "burned in my memory," for he had realized that, at that moment, Walker was saying that this was the last time he would see his best friend. Scott had not known Walker well, although he had often heard his father speak of him. On that day, his reaction was, "What a fatalist! But a true gentleman, and military all the way."[46]

In May 1942, Walker advanced to temporary colonel, then to colonel; and in June, to temporary brigadier general. He sent a copy of the press release to his sons, sounding pleased that "I'm going to be a field soldier after all" and joking that the Senate might balk at confirming his nomination. He expected to "receive orders for overseas and will have to leave within 48 hours thereafter. Am set up for a Bomber Command. Am told that the previous plans for my assignment have been changed and that now I'm going westward."

At that time, Marguerite and the boys lived in California. Walker planned, if possible, "to see my two fine fellas. I'm counting on that very very much."[47] He wrote Marguerite, asking if the boys could meet him to spend the afternoon. She wrote back not only agreeing, but inviting him to join them for dinner after their visit.[48]

He was busy, he wrote the boys, trying to "get my house in order." Perhaps joking, he told them it was "quite a task." He bundled up a lot of his clothes and sent them to his sons, writing them that cloth might get scarce during the next two years and their mother might be able to have them cut down to fit. He included his prize Peal boots and woolen riding breeches. Mary Hodge remembers that he was "pleased to be going but regretted going, too." She felt that his regret centered on leaving Cleo, "for they were very serious." Sometimes he joked, using the words "if I come back." Mary

Hodge did not think he dwelt on his possible death, but she sensed that he felt he would not return. She dismissed it as the usual reaction of a man going off to war.[49]

The Hodges offered what support they could, and Walker brought his personal papers to their house to store on the basement bookshelves. Mary looked through them enough to realize that here was his entire professional life—notebooks and more notebooks, mostly written in pencil, full of sketches and designs of planes. He talked a little about wanting to write about his work if he came back. Not only did he not live to do so, but his papers also did not survive. After the war, Hodge was named ambassador to Japan and the Hodges rented their house while overseas. Needing space, the tenants destroyed Walker's materials. The loss of his personal papers means that Kenneth Walker had almost no opportunity to speak for himself to historians, to reveal his beliefs and doubts and convictions, to explain his reasoning, to answer criticism. The unfortunate result is that while we can be certain of many things he did, we can be certain of much less about the man himself.

On 17 June, Walker made out his will, leaving his estate to his three sons. On the 19th, his name was on a list of officers assigned to the Allied Air Forces in Australia. His orders came through on the 22d.[50]

The night before he left Washington, he had dinner with the Hodges and they drank champagne to his return. To Mary Hodge, he left the image of a man "who knew his profession and knew it to perfection, a man who wanted things done this way and not that way, and they should be done yesterday instead of tomorrow. He was sometimes very cross with the way things were done—he knew they should be done differently and he would have done them differently." Asked if Walker was intolerant, she responded, "I wouldn't say he was intolerant, but that's just the way he was. He had two distinct personalities: everything had to be done just right, and he certainly was a military man from his toes to the top of his head."

But more than anything else, Mary Hodge remembered Ken Walker as "such a kind, pleasant person. We were all very, very fond of him."

Ken Jr. and Douglas met their father at the airport in Los Angeles. There had been a similar meeting when he came through Los Angeles en route from Hawaii to Washington. On that earlier occasion, the first time Douglas remembered his father at all, he wore a business suit and took his sons shopping and then to lunch. This time, he wore his uniform—first khakis, then the three went to his hotel where he changed into full dress. He bought a bouquet of flowers for Marguerite, then he and the boys boarded a trolley to her home in Glendale. "You can imagine the immense pride I experienced," Douglas later wrote, "as the passengers on the crowded trolley began to realize there was a general officer standing among them." Only nine years old, Douglas also felt "an odd sensation of verification. Here I was, like everyone I knew, with a father." He remembered little more about the visit, except that his father kissed his mother politely on the cheek, then stood in the kitchen door and talked to her as she finished preparing their meal.[51]

Kenneth Walker had done all he could do in the States. The next day, he boarded a plane and turned his face toward the Southwest Pacific.

Notes

1. DeWitt S. Copp, *Forged in Fire: Strategy and Decisions in the Air War Over Europe, 1940–45* (Garden City, N.Y.: Doubleday, 1980), 128.

2. Richard Aldworth to Lauchlin Currie, letter, 10 September 1941; and Currie Papers, Box 2, Hoover Institution.

3. Michael S. Sherry, *The Rise of American Air Power: The Creation of Armageddon* (New Haven: Yale University Press, 1987), 59, 76–79 (Bullitt quote 76).

4. Wesley Frank Craven and James Lea Cate, eds., *The Army Air Forces in World War II*, vol. 1, *Plans and Early Operations January 1939 to August 1942* (Chicago: University of Chicago Press, 1948–1958; new imprint, Washington, D.C.: Office of Air Force History, 1983) 104; and Assistant Chief of Staff, Intelligence (hereafter cited as Asst. Chief), *Air Action in the Papuan Campaign, 21 July 1942 to 23 January 1943*, USAF Historical Study 17 (Maxwell AFB, Ala.: USAF Historical Division, Air University Library Documents Section, M-U 27218, no. 17, 1945), 44–45.

5. Asst. Chief, chap. IX.

6. Ibid., 338–39.

7. Ibid., 375–80.

8. Charles E. Kirkpatrick, *An Unknown Future and A Doubtful Present: Writing the Victory Plan of 1941* (Washington, D.C.: Center of Military History, US Army, 1990), 77–78, 82–83, 90–91.

9. Asst. Chief, 44–45; Haywood S. Hansell Jr., *The Air Plan That Defeated Hitler* (Atlanta: Higgins-McArthur/Longino and Porter, 1972), 62–65; James C. Gaston, *Planning the American Air War: Four Men and Nine Days in 1941, an Inside Narrative* (Washington, D. C.: National Defense University Press), 12–4; and Memorandum for the Chief of Air Staff, subject: Notes on Preparation of AWPD-l, 18 November 1941, 145.96-154, USAFHRC.

10. Ethel Kuter, telephone interview with author, 1991.

11. Gaston, 38–39.

12. Hansell, 65.

13. Gaston, 14–15.

14. Other officers working on AWPD-l included Lt Cols Max F. Schneider and Arthur W. Vanaman, Majs Hoyt S. Vandenberg and Samuel E. Anderson. Robert Frank Futrell, *Ideas, Concepts, Doctrine: A History of Basic thinking in the United States Air Force, 1907–1964* (Maxwell AFB, Ala.: Air University, 1971), 59–62.

15. Asst. Chief, 375-80; Wesley Frank Craven and James Lea Cate, eds., *The Army Air Forces in World War II*, vol. 1, *Plans and Early Operations January 1939 to August 1942* (Chicago: University of Chicago Press, 1948), 137–38, 559.

16. AWPD-l Scrapbook, 145.82-1, USAFHRC; Hansell, 75.

17. Kuter, unpublished manuscript, US Air Force Academy Library, Special Collections, 155.

18. Hansell, 76–77.

19. AWPD-l Scrapbook; Hansell, 78–85.

20. AWPD-l Scrapbook.

21. Gaston, 22.

22. Hansell, 88; Craven and Cate, vol. 1, 131–32, 145–50.

23. Kuter, interview, 203–4.

24. AWPD-1 Scrapbook.

25. ACTS Lectures, "The Use of Aviation in the Spanish War" and "Aerial Warfare in Spain," cited in Thomas A. Fabyanic, "A Critique of United States Air War Planning, 1941–1944" (PhD diss., Saint Louis University, 1973), 66–67.

26. Futrell, 52; Haywood S. Hansell Jr., "Harold L. George: Apostle of Air Power," in *Makers of the United States Air Force*, ed. John L. Frisbee (Washington, D.C.: Office of Air Force History, 1987), 82; and Fabyanic, 69–71.

27. Laurence S. Kuter (unpublished manuscript, USAFA Library Special Collections), 133, 158.

28. Maj Haywood S. Hansell Jr., memorandum to Gen Henry H. Arnold, subject: An Air Estimate for the Situation for the Employment of the Air Striking Force in Europe (ABC-1), 11 August 1941, 9–10, cited in Fabyanic, 73–74.

29. Hansell, *The Air Plan*, 107. A special board, assembled in October 1941, to study the pursuit problem, recommended development of high- and low-altitude interceptors, night fighters, and long-range, multiplace fighter escorts, with the lowest priority given to escorts. Futrell, 59.

30. Fabyanic, 64–65, citing AWPD-l, part 2, Tab 3, 2–3; Part 3, Tab 35, 1; Craven and Cate, vol. 1, 149, citing Tab 4 of AWPD-l; and Hansell, *The Air Plan*, 92.

31. Gaston, 76; Fabyanic, 64; and Mary Hodge, interview with author, Arlington, Va.

32. Gaston, 81; Hansell, 90.

33. Memorandum for the Chief of Air Staff, subject: Notes on Preparation of AWPD-l, 18 November 1941, 145.96-154, USAFHRC. Lt Col Arnold George memorandum to Chief of the Air Staff, subject: Information Reference AWPD-l, 15 September 1941, 145.82-1, USAFHRC.

34. AWPD-1, Munitions requirements of the Army Air Forces to defeat our potential enemies, 12 August 1941. Includes revisions, etc. 145.82-1, USAFHRC.

35. Memorandum, 25 September 1941, 145.82-1, USAFHRC.

36. Quote as given in Copp, *Forged in Fire*, 174.

37. Ibid., 194.

38. Craven and Cate, vol. 1, 236, 246; and Futrell, 63–64.

39. The identity of the woman here called "Cleo" is hazy. Walker's sons knew of her but do not remember her name. Mary Hodge and her daughter, the primary sources for this portion of Walker's life, provided the name "Cleo" but professed not to remember a surname.

40. Mary Hodge, interview with author, 16 October 1990.

41. Mary Hodge, Arlington, Va., and Mary Lee Barton, interviews with author, 16 October 1990.

42. Mamie Lee Andrews, Chatham, Mass., and Ethel Kuter, San Antonio, Tex., interviews with author.

43. Walker to Ken, letter, 1 March 1942.

44. Walker 201 file; citation in K141.2421, USAFHRC.

45. Walker to Ken, letter, 1 May 1942.

46. Scott Pearson, Denver, Colo., telephone interview with author, 7 February 1991.

47. Walker to sons, letter, 15 June 1942.

48. Marguerite Walker, videotape interview.

49. Walker to Ken and Doug, letter, 21 June 1942; and Mary Hodge interview.

50. Copy of will in possession of Douglas Walker. AG210.311, 19 June 1942.

51. Douglas Walker, New York, interview with author, 19 January 1992.

Chapter 5

The Southwest Pacific and Fifth Bomber Command

Flying west and south toward Australia, Walker knew that he faced a combat challenge of considerable magnitude. The Japanese armed forces had dealt the Allies a series of devastating defeats in accomplishing exactly what they had set out to do: gain control of the resource-rich lands around the South China Sea, isolate China, and extend Japan's conquests into the Indian and Pacific Oceans to provide a defense line in front of their conquered areas. By June 1942, that defense line included Wake and the Marshall Islands to the east, Burma and the Netherlands East Indies to the west, and—most disturbing—New Britain and much of New Guinea to the south. Rabaul, a fine natural harbor on New Britain, had been taken from the Australians and turned into a major naval and air base. From Rabaul, the Japanese were moving down the chain of the Solomon Islands and onto the Papuan Peninsula of New Guinea. Just across the Coral Sea, vulnerable and apparently next in line of conquest, lay Australia, the kingpin of Allied Pacific strategy. Distances across the Pacific were so vast that if Australia were lost as a forward base, few would venture to guess how long the war might last.

Gen Douglas MacArthur, Supreme Commander of the Southwest Pacific Area, established his headquarters in Australia in March 1942 after leaving the doomed Philippines. Australian forces stationed in North Africa and the Middle East were rushed home for Australia's defense. Allied success in the battles of the Coral Sea (3–8 May) and Midway (4–6 June) blunted the Japanese advance and bought a little time. Plans were underway to seize the initiative, but the Southwest Pacific was still touch-and-go.

Walker flew to Australia in company with Brig Gen Enis C. Whitehead, who was shocked by the confusion and lack of organization they found on arrival.[1] The combined Australian and US air forces were in no shape for combat. MacArthur had little confidence in his air forces and even less use for their commander, Lt Gen George H. Brett. Maj Gen George C. Kenney was scheduled to replace Brett, but he would not arrive for another month. Caught in a situation in which he could do little, Brett sent the two new generals on an inspection trip so they could see for themselves the conditions they faced. He asked Walker to make a complete study of the existing bombardment methods. They were not being handled as effectively as possible, he feared, and he urged Walker to see if he could not put more pep into it.[2]

Walker learned a great deal during his trip. He visited the 19th Bombardment Group, battered veterans of the Philippine and Java campaigns, at Townsville. He observed that the B-17s could fight off the Zeros when necessary, for gunners had "plenty of guts and make the Japs pay for any attack they launch."[3]

One day he joined a B-17 crew as an observer on an unaccompanied reconnaissance mission to search for a convoy. They found it while at 6,000 feet, then circled to get 10,000 feet altitude. "Fortunately, there were no Zeros around," Walker wrote, "although I was foolishly disappointed for a while." After antiaircraft opened up, he got quite a kick out of watching the guns flash and a few seconds later seeing puffs right below us or to the sides. "Shell fragments sounded like hail on the wings, and we got one fair-sized hole in the right wing. It was my first time under fire, but I was so interested that I forgot to feel concerned."[4]

He flew on two other combat missions with the bombers and was impressed with the challenges they faced. A mission against Japanese positions on the northern coast of New Guinea, one of the most primitive and inaccessible areas of the world, could last 16 to 18 hours over hostile ocean, mountain, and jungle. Maps were insufficient and inaccurate, intelligence almost nonexistent, distances vast. From Townsville to the Japanese base at Rabaul was 1,095 miles. The tropical weather seemed determined to humble mere men and planes; storms could cause such severe turbulence that crews might return with broken bones. Cumulus clouds began forming by midmorning to tower over the ocean and the dark mass of New Guinea's Owen Stanley Mountains—mountains that sent vertical shards to heights of 13,000 feet. Those who survived a crash had little chance for rescue or escape. The swamps and jungles swallowed a downed plane like a frog gulping flies. This was primitive, forbidding country. The natives were described by Walker as black, bushy-haired, rather small, "and they smell." Further, they were known to go cannibal once in a while. Obviously, the Japanese would be only one of the hazards facing Walker and his men.

Port Moresby, on the southern side of New Guinea's Papuan Peninsula, figured prominently in the scheme for future operations. During that summer of 1942, it was used primarily as a staging point for the bombers based in Australia—for the Japanese bombed Moresby steadily. But Moresby's harbor was large enough to shelter a fleet, and the town lay in a coastal plain where airfields were feasible. New fields were under construction, with grass huts for alert shacks. The men lived in tents and fought mosquitoes, mold, and mud during the rainy season, pervasive winds and dust during the dry season. Malaria, dysentery, and other diseases took a heavy toll.

While at Moresby, Walker found out how it felt to be on the receiving end of an air raid. "The bombs fell pretty heavily and pretty close," he wrote. "I crawled into a slit trench with the rest of the people—feeling foolish. One quickly gets over that feeling, however."[5]

Walker's initial impressions from his survey were mixed. He was unimpressed by Australia as a land, describing it as "a little drab." The men stationed at Torrens Creek lived in a semidesert of red dirt with little water. Those at Fenton Field might see kangaroos and wild horses and enjoy good hunting for recreation, but the land was so bleak and raw that men could easily get lost. Iron Range was little more than a field called out of the jungle, and the water supply at Horn Island was precarious. Great distances combined with limited transportation would put extra demands on the air forces, and there were few on hand. US air strength in the theater totaled only 1,602 officers and 18,116 men. On paper, there were three fighter groups and five bombardment groups, supported by two transport squadrons and one photographic squadron. The 8th (P-39s) and 49th (P-40s) fighter groups in Australia were at strength and the 35th (P-40s) at Port Moresby was at partial strength. Of the bomber groups, however, the 43d had no planes, while the 19th (H) carried physical and moral scars from the devastating campaigns in the Philippines and Java. The 22d (M) with B-26s was operational, but the 38th (M) did not have its B-25s in commission. The 3d Bombardment Group (L) was equipped with an assortment of planes, including B-25s, A-20s, and A-24s. High humidity and rough conditions placed extra demands on maintenance crews while replacements and parts were limited by shortages in both planes and transport.[6]

Walker's assessment of the men, however, was positive. Visiting with a pursuit unit in Australia's northwest, he was favorably impressed with the men and their record—60 Japanese planes downed for a loss of five of their own. The pilots, he noted, were young kids; but one had 10 kills to his credit and two others had eight each. Listening to the pilots, he had no doubt that "our boys will lick 'em."[7]

"These young pilots are plenty brave," he wrote. "Just had dinner with a 2d Lt who is just out of the hospital. A Zero got on his tail and shot him down. He bailed out at around 500 feet. Pulled out of a dive of about 500 mph and managed to get out."[8]

Early in August, Walker had an opportunity to see even more of the country. A civilian from Washington came to Australia on a classified mission to garner information about uranium deposits. Whether or not Walker knew the scope and direction of atomic research, he accompanied the civilian on his trip. A B-17 picked them up at Townsville on 4 August and flew them first to Darwin, where they were met by a party of Australian officials. On 6 August they went on to Alice Springs, in the center of the vast country, again being met by Australian officials. After a few hours at Alice Springs, they returned to Townsville that night. During the trip Walker spent a lot of time in the cockpit with the pilot, who noted that "the general was extremely interested in our bombing operations."[9]

General Kenney reached the theater on 28 July. He had already cleared the air with MacArthur and taken hold of the air forces by the time Walker returned. Walker and Kenney had served together at the Air Corps Tactical School, where Kenney's strong personality and innovative thinking had left a legacy that matched Walker's for forcefulness. Among other assignments, Kenney had been chief of the production engineering section of the Air Corps Materiel Division and commander of the Fourth Air Force. "He will make a splendid commander and I'm proud to serve under him," Walker had said.[10] Kenney, for his part, considered himself lucky to have both Walker and Whitehead, men who had brains, leadership, and loyalty, and who liked to work.[11]

Brett had figured Kenney would want Walker as a staff officer, but by the night of 5 August Kenney had decided that Whitehead would for the time being go to Port Moresby as advanced echelon commander, handling all operations and giving final instructions to the bombers coming in from Australia. Instructions for bomber strikes would be transmitted from Brisbane (Kenney's headquarters) to Walker at Townsville. Walker would command the Allied Air Forces in the Northeast with the help of Group Captain Garing of the Royal Australian Air Force (RAAF). RAAF combat units in the Northeast—Townsville, Cairns, Horn Island—would be attached to Walker for operational control. Later, Walker would be appointed bomber commander. For now, Kenney determined that Walker would set up the missions "in accordance with my instructions," brief the crews and dispatch them to Port Moresby for topping off.[12] Obviously, Kenney intended to take an active role in operational decisions and keep his commanders on a tight rein.

During these first weeks, however, Kenney gave Walker authority to take charge at Townsville and reorganize the Northeast Area Command, which had impressed Kenney as a "scrambled outfit of Australians and Americans [that] resembled a can of worms." Before Kenney arrived, it had already been agreed that Americans and Australians would be separated organizationally; the Australians would assume responsibility for their country's defense and free the US forces to take the offensive. The total reorganization took time, but the US Fifth Air Force, with Kenney as commanding officer and Walker heading the Fifth Bomber Command, officially came into being on 3 September.[13]

Generals Kenney and Walker

Meanwhile, the military situation called for haste. The Allies had planned to establish airfields at Buna, on the northern shore of the Papuan Peninsula. Unfortunately, the Japanese landed troops and set up their own base at Buna in late July. Japanese troops, apparently intent on capturing Moresby itself, began marching southward from Buna via the Kokoda Trail over the Owen Stanleys. Australian troops advanced northward to meet them. At the eastern end of the peninsula, Allied engineers were hastening to build airfields at Milne Bay because planes based there could control the surrounding seas. The battle to hold Australia was rapidly taking shape on the Papuan Peninsula and in the nearby Solomon Islands, where on 7 August US Marines landed at Guadalcanal. Both battles promised to be grim, with the outlook by no means assured. The air forces in Australia, whatever their limitations and problems, must provide support for both operations.

Kenney wasted no time. Those he considered deadwood, he transferred. He set Whitehead to work pushing construction on airfields and revetments in New Guinea. He wanted the bombers to move forward and intensify their operations. To correct the dreadful disorder in maintenance and supply, he started construction on a central depot at Townsville. Perceiving that the war-worn 19th Bomb Group was "shot," he began plans to replace it with the 90th.

Kenney took up what he called the bombing mess with Walker. Japanese fighters were interfering with most missions—and the bombers didn't hit anything when intercepted. He noted grimly in his diary: "Our own short-sightedness, mine included, didn't put the range in our fighters to do this job out here." Walker's side of this conversation is not on record, but obviously he was not as ready as Kenney to accept the bomber's vulnerability. He did not react with enthusiasm to Kenney's plans, which included starting work at once on low-altitude operations. It would be some time before P-38s could replace existing fighters and provide escort to distant targets. Meanwhile, Kenney expected low-altitude tactics to result in less trouble from fighters and more bomb hits. The Navy in the South Pacific was experimenting with dive bombing; and Kenney was eager to work on a new tactic called skip bombing.[14] Such measures take time, and the Japanese continued to advance. Late in August, they landed at Milne Bay despite efforts by the B-17s to destroy them at sea.

The fighting at Milne Bay lasted for several weeks before the Allies gained control; and during that time, the Japanese march across the Owen Stanleys sent the Australians into retreat. By 12 August, the Japanese held Kokoda and its small airstrip; by the end of the month, they were less than 30 miles from Port Moresby. Australian reinforcements were flown in, joined in mid-September by part of the US 32d Division. At Kenney's suggestion, one regiment was flown from Australia to Moresby, the aerial transfer taking less time than by water. The previously skeptical MacArthur suddenly became an enthusiastic convert to airpower, convinced that air can do anything.[15]

The campaign in New Guinea, where there were neither roads nor railroads, began taking shape with the air forces putting down covering fire, transporting troops to the combat zone, and keeping them supplied. They used airdrops until landing fields could be carved from the jungle. But this all-encompassing application of airpower had received little attention. "There is nothing in the book about it," Kenney observed, calling it a proving ground where you have to invent your methods and tactics as you go along.[16]

Walker enthusiastically supported this broad use of airpower to advance the ground forces.[17] But he was hard-pressed to meet all the demands on bombardment—strikes against shipping, strikes on the main base at Rabaul plus Japanese positions in northern New Guinea, and also supplementing the limited transport units in supplying the troops. Bombers also had to be diverted to reconnaissance duties because of the great distances involved. Crews were overextended. "The A-20 boys are getting worn down," Walker warned Kenney on 7 September, but Kenney figured that Moresby would be lost in a week without their efforts.[18]

Shortage of planes as well as crews limited what Walker could do. In September, for instance, he had the available B-17s loaded with parafrag bombs to attack Japanese fields at Lae and Salamaua, on the northern cost of New Guinea. Kenney had ordered this mission to slow down the advance on Moresby. But a crisis arose at Guadalcanal, a convoy of reinforcements had to be run in, and a supporting strike on Rabaul was requested to keep the Japanese planes diverted. The B-17s had to be unloaded and prepared for a different mission with different bombs.[19]

In mid-September, when the Papuan crisis was at its worst, Kenney ordered Walker to Moresby for several weeks to direct the advanced echelon, a move calculated to give Whitehead a rest and Walker more experience. Living conditions in New

Guinea were basic, with few personal amenities and unappetizing rations. Eddie Rickenbacker, who came through Moresby about this time on a Pacific inspection trip, called Moresby the dust bowl of all creation, and when Kenney and MacArthur joined Walker for lunch one day, Kenney observed that it was the worst mess in New Guinea.[20]

Walker took it all in stride and did what he could to improve morale. In December, after Fifth Bomber Command headquarters moved to Moresby, clerk William Travis was impressed that Walker not only was the best typist in Fifth Bomber Command, not above pounding something out for himself when things got tight, but that the general dealt with the enlisted men on a person-to-person basis, wore open-necked khakis, stood in chow lines, and treated the men as equal human beings. In retrospect, the clerk considered Walker "the best soldier I ever knew, from every point of view. Even without the externals of rank, you knew he was the general."[21]

By the end of September, the critical phase had passed and the fighting along the Kokoda Trail was turning in the Allies' favor. A relieved MacArthur radioed Washington that, with "the energetic support of his two fine field commanders, Whitehead and Walker," Kenney had vitalized the air forces and raised them "from unsatisfactory to very good, and soon to excellent."[22] Morale was definitely on the rise. Bomber crews had begun to feel a sense of purpose and direction, a change some of them attributed to Walker's leadership. Before he came, one crew member remembered, "seems like we never understood where orders were coming from or why we were doing anything—it was a catch-as-catch-can sort of thing." By September, however, "we started doing things."[23]

Part of the rise in morale resulted from Walker's practice of accompanying the bombers on their missions. "He figures he can't direct flights from the ground and tell the boys what they are doing wrong," his aide and pilot, Capt Fred P. Dollenburg, explained to newsmen. "So he goes along and directs a flight from the air. If a plane gets out of formation, he shouts his orders over the radio to 'get the hell back in line.'" Walker's presence not only sharpened tactics but boosted morale and raised him in the estimation of his men. "The general figures he can't tell the boys how to go out and to get shot at unless he's willing to get shot at too," Dollenburg said.[24] When Walker was awarded the Silver Star in August, the citation noted that the large amount of firsthand information gained had been of "inestimable value," while his action "proved highly stimulating to the morale of all Air Force personnel with whom he has come in contact."[25]

"He put his life on the line," Travis said simply. "He wouldn't send the bombers out on missions and enjoy the fruits of their labors without sharing the dangers as well."[26]

General Arnold, Chief of the Air Force, visited the theater at the end of September. He decided Kenney was "a real leader and has the finest bunch of pilots I have seen." After he and Walker discussed personnel problems over breakfast, he promised to find a way to exchange the weary 19th Group or send in 10 combat crews per month as replacements. From MacArthur (in Brisbane), he heard only praise for Kenney, Whitehead, Walker, the men, and even the planes. Arnold left the theater convinced it was time for the Allies to take the offensive. "If we don't," he noted tersely, "the Japs will."[27]

Despite MacArthur's blanket praise, the performance of the bombers had been, and continued to be, disappointing. A Joint Chiefs of Staff (JCS) officer (who had visited New Guinea in September) reported, "Our bombardiers could not hit anything from altitude." Admitting that bombing results left much to be desired, Walker stressed training. He also spoke up for the need to tailor strategic plans to the means at their disposal, for there were more targets and needs than there were planes and crews to meet them. He had tackled the complex problem with his customary energy, and a visiting JCS officer noted that corrective measures in both training and tactics had been vigorously initiated.[28]

Part of the vigor that was precipitating change came from Kenney. Before long, he and Walker were locked in a tense conflict. Both generals were sharp and creative, outspoken and high strung, and blessed with strong, healthy egos. They did not see eye-to-eye on what to do and, like any officer caught in a disagreement with his superior, Walker became testy. Never hesitant to act, Kenney began implementing weapon and tactical changes as soon as he reached the theater. When Kenney first suggested on 10 August that the bombers use low-level attacks and instantaneous (rather than delayed) fuses, Walker objected. He had built his career around high-altitude formation bombing. Kenney's point of view was that this made an excellent approach for big targets like cities or airfields, but Fifth Bomber Command targets were frequently ships—not only smaller targets than airfields, but targets in motion. For the moment, however, Kenney let Walker have his way and dropped the matter.

Walker was working on improving formation attack through changes in the approach and bomb run, using what he learned on the actual raids as a guide. As described by Donald Wilson, now a brigadier general and Kenney's chief of staff, Walker "developed a scheme where the bombardment formation would split into three elements and approach the target from three different directions, thereby intending to confuse the antiaircraft and other defenses, and coordinate their bombing attack by timing and careful flying of their courses."[29]

Walker had his mind set on future mass raids against Rabaul; but in the fall of 1942, Fifth Bomber Command missions seldom numbered more than five or six planes, hardly enough for the minimum pattern bombing. The pugnacious Kenney wanted results now, and he wanted his ideas tested. On 19 September, he ordered Walker to give instantaneous fuses a try against shipping. "Ken didn't like the idea," Kenney wrote, "and his Naval liaison officer didn't think much of it, either, but I told them to try it for a while and see what results we got."[30]

A critical military situation on both Papua and Guadalcanal heightened the tension and, on 5 October, Kenney got his dander up over the lack of success in the previous day's bombing missions. Six B-17s had bombed antiaircraft batteries at Buna, six B-25s attacked a Japanese convoy approaching Buna but scored no hits, and 11 B-17s of the 19th Group attacked Rabaul. "Reports show formation did not hold," Kenney noted grimly. "Wrote Walker and told him to stop piecemeal attacks."

Kenney also added a warning that Walker, who reportedly had flown on the Rabaul mission, must stay out of combat. "I can always hire a 10 dollar a week man to sweep the floors. No more combat missions."[31]

Three days later, Walker called off a scheduled strike against Rabaul on the basis of a bad weather forecast. Kenney had a lot riding on the mission, for MacArthur had asked him to conduct it in support of the hard-pressed forces at Guadalcanal. The Japanese had been pouring men and supplies into Rabaul for several weeks. Since they unloaded at night and left before dawn, their ships were seldom damaged by US raids. Kenney had decided to slow things down by burning out the town of Rabaul with its supporting facilities. When he learned that Walker had canceled the first of the two planned strikes, Kenney consulted a different weatherman, got a favorable forecast, and overruled his bomber commander.

The two Rabaul raids flew as scheduled, instantaneous-fused bombs proved effective, and the raids were considered highly successful. One of the squadrons taking part was the newly formed 63d, under Maj William G. Benn, a promising young officer whom Kenney had brought to the theater as his aide. Kenney noted with pleasure that the 63d had already become the hottest outfit in the whole Air Force.[32]

A few days later, when Walker saw Kenney in Brisbane, Kenney noted that his commander's "feelings were hurt because I had countermanded his orders canceling the Rabaul show on the 8th. I kidded him." No doubt Walker seethed under Kenney's teasing, but when Kenney explained the particular pressure from MacArthur to conduct the raids, it took some of the edge off his resentment. "Walker OK now," Kenney wrote in his diary—"feels much better."[33]

Walker's reluctance to go along with Kenney's suggestions contrasts with his reaction to other approaches for improving combat efficiency. Individual initiative and imagination were encouraged in the Fifth Bomber Command, with free exchange of ideas between units and the higher command.[34] Patrick Norton, for example, in 1942 a buck sergeant in the 22d Group, had made some modifications in his B-26 that improved its performance. Walker heard about it and went to take a look. He considered the idea sound, praised Norton, and took the necessary steps to ensure that new models would carry the improvement. Norton formed the impression

Unidentified; General MacArthur; General Blamey; General Kenney; Unidentified; General Walker

General Walker; Navy Secretary Knox; Unidentified

of a commanding officer "trying very hard to do a good job without nearly enough equipment, and anything that made the equipment more efficient or effective was very important to him." The command might have teething troubles, but the men and their commander were working well together.

Norton's positive impression of Walker was reinforced in mid-September. His modified B-26 was shot down and demolished, the copilot was killed, and Norton himself ended up in the hospital at Marabou, a base north of Townsville. Needing some information about the crash, Walker came to see him. After Norton supplied the requested information, Walker said, "I know how proud you boys were of your airplane. Hurry up and get well, and I'll try to have another airplane for you when you get out." Blaming his reaction on the situation—"I was banged up pretty good and about half doped up"—Norton fired back, "Who in hell told you that I wanted another plane?" Rather than reprimand him for speaking out, Walker grinned, turned to the nurse, and said, "I think I've worn out my welcome."[35]

In his relations with Kenney, Walker was more defensive. Kenney had come to the theater fired up with the possibilities of low-altitude skip bombing. The Royal Air Force had used the tactic effectively; the US Armament School at Eglin had also done some work with the technique. As early as 13 August, Kenney encouraged Benn, who shared his enthusiasm, to start experimenting. The first trial run was made on a sandbar. After Benn was given command of the 63d Squadron, he and his crews practiced against a wrecked ship in Port Moresby's harbor. Approaching their target at 200 mph, they released their bombs about 300 yards from the wreck at 200 feet or lower. When speed, range, and altitude were coordinated, the bombs would skip across the water into the side of the ship.

Toward the end of October, Benn reported that he was ready to try skip bombing against a real target. Kenney told him to go ahead. Over Rabaul harbor on the night of 23 October, six B-17s bombed from altitude, then six B-17s from the 63d Squadron dropped to 100 feet and skipped their bombs into the Japanese ships. They claimed to have sunk *several* smaller vessels *plus a* destroyer that was credited to one of the 63d's captains, Ken McCullar.[36]

Kenney was delighted. "This was the first skip-bombing for keeps," he wrote. "Walker still is not keen about skip-bombing and does not like the fact that Benn talks to me about it. I tell Walker that Benn and I have been fooling with this thing for a long time and not to worry about it. If it is good we will do something with it." He hoped this successful trial would "quiet Walker's fears."

Walker's fears probably had their roots in concern for his crews. In the debate over level- versus dive-bombing, he defended the former on the grounds that dive-bombing cost too heavily. Even Kenney worried that the B-17 did not have enough forward machine guns to fly into a ship's defenses at low level, and for that reason he had instructed Benn to make his attack by night.[37]

By this time, Walker and Benn had already clashed. "Found out that Walker has been giving Bill Benn the devil for not obeying orders," Kenney wrote on 15 October after he and Walker had held decorating ceremonies at Townsville and Mareeba. "Walker decided after a couple of missions to go back to the 1/10 second delay fuze instead of instantaneous fuze settings on shipping attacks. Benn knew that I wanted the instantaneous fuze used so he kept on using it." Choosing to consider this "a clash of personalities" rather than disobedience of orders, Kenney was determined "to keep them together somehow." Walker was the Army's leading bombardment expert, and Kenney already intended to give Benn command of the 43d Bomb Group after he had gained more experience. Kenney settled the immediate issue by telling Benn he must obey Walker's orders, then telling Walker that "I wanted my idea given a month's test and that if I heard of any more 1/10 second delay fuzes being used on shipping attack I would take away from the 5th Bomber command the privilege of deciding on their fuze settings. Ken said 'Yes sir' with no more argument."[38]

According to an unverified story, probably based on this 15 October confrontation, during one argument between the sparring generals Kenney pulled rank, whereupon Walker saluted and said, "Okay, but f _ _ _ you, George."[39]

After some weeks of using instantaneous fuses, Walker on 18 November asked Kenney for permission to go back to the 1/10 second delay fuse. Walker's reasons for making the request are not in the record, but Kenney was armed with statistics from the preceding month's trial. Analysis showed convoys slowing down after being attacked, with more ships being sunk and many more reported damaged and on fire. Kenney decided he would reinforce this data with a demonstration.

"I told Ken to have somebody go out and drop about four bombs at the old wreck on the reef outside Moresby," Kenney wrote, "and that we would then go out and inspect it."

The bombs were dropped; a motorboat took the generals within a mile of the wreck; a corporal rowed them the remainder of the way. The evidence proved Kenney right. The bombs had missed the vessel 25 to 75 yards and yet fragments had torn holes all through it. Some of them were two to four square feet in area. Sharp, clean edges distinguished that day's holes from others made earlier.

Walker could only concede gracefully. "Okay," he said. "You win. I'm convinced." Kenney then ordered the corporal to sit in the stern while Walker rowed them back to the motor boat. Kenney may have been rubbing his victory in, but Walker rowed in silence, voicing only "a few three- or four-letter [words] when his oarsmanship went wrong."

Kenney was big enough to let the matter drop at that point. After a few drinks that evening, Walker "thawed out," as Kenney put it. "Ken is O.K.," he wrote. "Stubborn, over-sensitive and a prima donna but works like a dog." He noted that Walker's men liked him but he had trouble delegating authority. "I'm afraid Ken is not durable enough to last very long under the high tension of this show," Kenney noted. "His personal problem is tough because he keeps himself keyed up all the time and can't seem to relax a minute." Kenney began thinking of sending Walker back to Washington, where he would be valuable on Arnold's planning staff.[40]

Throughout November and December, the military pressure continued unabated. On the Papuan Peninsula, Allied ground forces, with unprecedented assistance from air forces for transport, supply, and military support, drove the Japanese back to Buna and began the campaign to secure New Guinea. Walker's Bomber Command was hard-pressed. The problems seemed endless. Rain made airfields dangerous for takeoff. After the exhausted 19th Bomb Group went home, the 90th (which replaced it) was slow getting into action. Once tested in action, the crews proved painfully green and had to be pulled back for more training. From the beginning, the theater had been short of men. Despite efforts to utilize all possible trained personnel, including civilians, they never had as many as needed. Killed and missing ran about 40 per month; another 200 were constantly on the sick list from wounds or disease. In December, Walker questioned the wisdom of moving the US Army out of Australia to forward bases, noting that the RAAF did not have adequate personnel to keep its present units going and was not in position to expand.[41]

During these trying weeks, Walker continued to hold the admiration of his men. The 43d Bomb Group renamed itself Ken's Men, a gesture that Kenney believed was made in his honor, although in later years it developed that some members of the group believed it to have honored Walker while others thought it was in tribute to Ken McCullar, a master of the low-altitude attack. Today, the group considers its name a memorial to all three.[42]

Late one night, William Travis picked Walker up at Seven-Mile Field (the airfields at Moresby were named for their distance from town) and drove him back to the base. Possibly driving too fast, certainly feeling big about driving the general, Travis met an oncoming vehicle at a narrow bridge, could not slow down in time, had to squeeze by, and afterward nearly ran off the road. Realizing that he had scared the fool out of the general, Travis braced himself for a dressing down. Walker said nothing. "He fitted my idea of the gentlemen," Travis wrote. "He understood quite well that I felt miserable about it, and he was willing to overlook it without comment."[43]

Shortly after Christmas, we have another glimpse of Walker through the eyes of Patrick Norton. He and several buddies were looking for a ride back to their base at Iron Range. Hearing that a B-24 was getting ready to take off for Moresby, Norton, not realizing it was Walker's plane, asked the pilot if he would consider dropping them off at Iron Range en route. It meant some 200 miles of extra flight. The pilot said he didn't think the general would want to do this, but Walker overheard and interjected that he would be glad to take the men. During the flight, he sat beside Norton and carried on a relaxed, comfortable conversation. They discussed the alterations Norton had made on the B-26, then Walker started talking about other things being done. The 20-year-old Norton heard about innovations that seemed to him pretty remote and fantastic: "an airplane without a propeller," a new system that would enable one gunner to operate all a plane's guns, proximity fuses that would go off when they got close to the target and thus eliminate the need for a direct hit, radar to give advance knowledge, and heat-seeking bullets that would seek out a match in the sky.

"He talked about how there were really a lot of wonderful things happening," Norton remembered, "and what we really had to do was do as good a job as we could now, with the equipment we had, because there was more and better equipment coming." A week later, Norton learned that Walker was missing in action. He felt a deep sense of loss. The general, in his estimation, was "a thoroughly nice, decent guy."[44]

Notes

1. Maj Gen Enis Whitehead, interview, 30 May 1945, Tab 57 in Papers of W.D. Edmonds, 168.7022-6, US Air Force Historical Research Center (USAFHRC).

2. Comments of Brett on Personnel, Kenney Papers, vol. 1, Office of Air Force History (OAFH).

3. Kenneth N. Walker to sons, letter, 11 August 1942.

4. Ibid.

5. Ibid.

6. Wesley Frank Craven and James Lea Cate, eds., *The Army Air Forces in World War II*, vol. 4, *The Pacific: Guadalcanal to Saipan August 1942 to July 1944* (1950; new imprint, Chicago: University of Chicago Press, 1948–1958), 7–8.

7. Walker to sons, letter, 11 August 1942.

8. Ibid.

9. Lt Col Robert J. Hughey, USAF, Retired, to author, 21 January 1991. Hughey, a captain in the 28th Squadron, 19th Bombardment Group, was the B-17 pilot on the trip.

10. Walker to sons, letter, 11 August 1942.

11. George C. Kenney, *General Kenney Reports: A Personal History of the Pacific War, With Maps by the Author* (New York: Duell, Sloan, and Pearce, 1949; reprint, Washington, D.C.: Office of Air Force History [OAFH], 1987), 11.

12. Kenney Papers, OAFH, 5, 9 August 1942. See Craven and Cate, vol. 4, 100.

13. Kenney, 41.

14. Kenney Papers, vol. 1, OAFH, 12, 13, 15 August 1942.

15. Kenney Papers, OAFH, 10 December 1942.

16. Ibid.

17. Craven and Cate, vol. 4, 117, citing General Headquarters letter and report not found.

18. Kenney Papers, OAFH, telephone conversation, 7 September 1942.

19. Kenney, 98–99.

20. "Pacific Mission," *Life* 14, no. 6 (8 February 1943): 97; and Kenney, 103.

21. William W. Travis, telephone interview with author, 4 August 1991.

22. Radio C-501, 16 September 1942, Historical Records Survey (Inventory) (HRS[I]) A-966, National Archives (NA).

23. Patrick H. Norton, telephone interview with author, August 1991.

24. Associated Press story, dateline 21 September 1942.

25. Walker's Medal of Honor Scrapbook, OAFH.

26. John Perakos and William Travis, telephone interviews with author, August 1991.

27. Diary of Henry H. Arnold, 25 September 1942, 168.65-42, USAFHRC.

28. Kenneth N. Walker to Willoughby, Miscellaneous Notes re Air Force, 22 September 1942, HRS(I) A-104, NA; and Radio 547-C, 21 September 1942.

29. Donald Wilson, oral history interview, Set #2, OAFH, 114.

30. Kenney, 108.

31. Kenney Papers, OAFH, 5 October 1942.

32. Kenney, 120-21; and Kenney Papers, OAFH, 8 October 1942.

33. Kenney Papers, 12 October 1942.

34. Fifth Bomber Command, vol. II, sec. III, 35, 732.01.

35. Patrick Norton, telephone interview with author, August 1991.

36. A later assessment indicates that no vessels were hit. See Craven and Cate, vol. 4, 112; John L. Frisbee, "Skip-Bombing Pioneer," *Air Force Magazine* 73, no. 12 (December 1990): 87.

37. Kenney Papers, 23 October 1942; HRS(I), 2113, 4433, NA.

38. Kenney Papers, OAFH, 15 October 1942.

39. Douglas P. Walker, interview with author, 19 January 1992.

40. Kenney Papers, OAFH, 18 November 1942; and Kenney, 143.

41. HRS(I) 1555, 1603, 2113, NA.

42. Kenney, 140; and Larry Hickey, 43d Bomb Group Historian, telephone interview with author, 28 November 1990.

43. William W. Travis to author, letter, 5 August 1991.

44. Patrick H. Norton, telephone interview with author, August 1991.

Chapter 6

Walker's Last Mission

Late in December 1942, the war in the Pacific reached a turning point. There would be nasty fighting ahead, but victory at Buna was assured, the struggle for Guadalcanal was coming to an end, Australia appeared secure, and the Allies were on the offensive. The Fifth Air Force, however, could see little reprieve. Every indication pointed to a bitterly contested fight for the remainder of New Guinea. The Japanese would most likely reinforce their garrisons at Lae and Salamaua on the Huon Gulf of northern New Guinea during coming weeks. They would also try to push the Australian ground forces out of Wau, inland from Salamaua.[1]

Rabaul, little damaged thus far, continued to be the vital hub for Japanese operations. Toward the end of December, Allied reconnaissance planes (which monitored Rabaul regularly) reported increasing concentrations of Japanese shipping in the harbor. By 30 December, this fleet had reached a peak to date—21 warships and an estimated 300,000 tons of merchant shipping. "When the Jap accumulates that much tonnage," Kenney wrote, "it means trouble for me shortly."[2]

After Fifth Bomber Command flew raids against Rabaul on 26 and 27 December, MacArthur advised the neighboring South Pacific theater that his available forces were temporarily exhausted. No new planes and crews had arrived to replace those lost in the steady operations of the preceding weeks. Adm William F. Halsey, South Pacific's commander, agreed to send his B-17s to Moresby to make possible a 10-day blitz against Rabaul. "Wish I had enough bombers to really go after them," Kenney fretted. Even with this combined force, the best he could manage would be one to three squadrons of heavy bombers every morning through 7 January.[3] Fifth Air Force had approximately 120 heavy bombers, but only 20 to 30 were available for daily combat. Some were held at Darwin for defense against the Japanese in the Netherlands East Indies, some were in depots undergoing repair or overhaul, some were in use for reconnaissance, and a percentage of the remainder were not risked at any one time.[4]

On 3 January, Kenney learned (through the top-secret Allied intelligence operation that had broken some of the enemy codes) that the Japanese were sending a convoy to Lae; it would leave Rabaul about 6 January. "Told Walker to intensify reccos on both N and S of New Britain," he wrote, "and put on a full-scale B-17 attack on Rabaul Harbor at dawn on the 5th to see if we can break it up at the source." To conserve strength and possibly achieve greater surprise, Kenney ordered no B-17 strikes to be conducted on 3 and 4 January, although Walker was told to "keep socking" Japanese fields at Lae and Gasmata (on New Britain) to prevent Japanese fighters from covering the expected convoy. And since the Fifth Air Force was going to need a forward airfield to support operations at Lae, Kenney ordered Walker and Whitehead to start looking for a site.

On 4 January, the Fifth bombed Lae, Sanananda, and Gasmata. Readying for the larger strike against Rabaul on the 5th, Walker told Kenney he wanted to hit Rabaul at noon rather than dawn as Kenney had ordered. Plans were for B-24s from Iron Range to rendezvous at Cape Hood with B-17s and B-24s flying from Moresby, and Walker was worried about their making a successful rendezvous if they took off at night for a dawn raid.[5]

According to the theory to which Walker had contributed so much, the larger and more compact the formation, the greater the likelihood of significant bomb damage and the greater protection the bombers could provide for each other. Thus far, however, Fifth Bomber Command had little experience with daylight formation bombing. Most of their operations had been either single-plane armed reconnaissance missions in daylight or small bombing raids by night, with only an occasional daylight formation attack on a convoy. Bombardment operations were still in their infancy, and Walker doubtless looked on this mission as an opportunity to test procedures. Hence his concern for rendezvous and formation.

Kenney, thinking of fighter opposition (no escort could be provided) and bombing results, continued to favor dawn. "The Nip fighters are never up at dawn," he wrote, "but at noon they will not only give our bombers hell but will ruin our bombing accuracy." As Kenney saw it, he would "rather have the kids not in formation for a dawn attack than in formation for a noon show." Noting that Walker was tired, jumpy, and showing strain, he nevertheless told Walker that he still wanted a dawn attack.[6]

Rabaul was a mean target—"a hot spot where you could get hurt pretty bad."[7] The Allied air forces had bombed intermittently since early 1942, the pace accelerating in the fall. Between 18 September and 30 November, in addition to raids in the Solomons, B-17s flew 180 sorties against Rabaul, B-24s flew two, and RAAF Catalinas, which often flew ahead to provide weather information or drop incendiaries to mark the target, flew 11.[8] Kenney noted grimly that these attacks brought results—"we are taking a heavy toll of Jap shipping and aircraft"—but the increased flying was costing planes, and "the Jap is getting replacements of his aircraft and we are not." Without replacements, his units had to "take it on the chin."[9]

The wear on the men was severe, for the raids on Rabaul were "terrifying and just plain miserable."[10] Wilson, Kenney's chief of staff, went on one of them as an observer in a B-24 and found it "quite an experience and quite a revelation." The town of Rabaul lay on the north shore of Simpson Harbour, a fine anchorage that opened into the larger Blanche Bay of St. George's Channel. Facilities were excellent: wharves, piers, hidden repair facilities, floating cranes, and provisions for submarines and small boats. The area stank of sulfur, for the harbor was a natural bowl surrounded by volcanic peaks. Nearby, to the south, lay the former Australian airfields of Lakunai and Vunakanau, which the Japanese had expanded and improved. Farther south were two additional fields, Rapopo and Tobera. Fighters, searchlights, and antiaircraft guns provided a full range of defense.

When the antiaircraft opened up, Wilson thought the scene below looked like an inferno, or possibly an active volcano. He noted with amazement that one plane flew low to interfere with the Japanese radar and listening devices, and he wondered "how in the hell an airplane could operate with all that shooting going on." At 10,000 or 12,000 feet over the target, "our experience with antiaircraft was to hear the boom of the explosions outside our aircraft and to feel the jolt of the concussion of the shells." Wilson watched the show from the side window. Later, the armament man asked if he wanted to go back into the bomb bay while he took care of the fuses on the bomb. Wilson did so, but when he saw what looked to him like a warehouse full of 500-pound bombs, each carrying 250 pounds of TNT that might be set off by a stray bullet, he admitted to "a squeamish feeling." One such experience seems to have satisfied him.[11]

The antiaircraft that so impressed Wilson was increased during the fall of 1942 (a total of seven 12-cm, twelve 8-cm, eight 7-cm, six 40-mm, and thirty-two 20- and 13-mm guns were added between August and December). On 1 January 1943, Col Nagaaki Kawai assumed command of an antiaircraft strength of seven battalions, an independent company, and five field machine cannon companies, plus three field searchlight battalions. General Imamura and the Japanese Eighth Area Army Headquarters moved to Rabaul in December 1942. The Japanese, under pressure from both New Guinea and the Solomons, were preparing for a major showdown.[12] Kenney perceived that he faced a tough enemy and that it would require "all our patriotism, stamina, guts and maybe some crusading spirit or religious fervor thrown in to beat him."[13]

Looking back, it is easy to see that Walker prepared for his last mission within a framework that offered only hard choices: marginal resources for offensive operations against a determined enemy, a major target that was defended well, and a conflict with his commanding officer over tactics. For reasons of his own, he defied Kenney's instructions, ordered the raid for noon, and accompanied it in person.

Exactly when he decided to fly on the mission is unclear. Officers of general rank were discouraged from taking part in combat for a number of reasons, a basic one being that their talents for high command were too valuable to risk. These same officers, however, often defied orders and went on combat missions—again for a number of reasons. It was an effective means of bolstering morale, it often provided insight that resulted in wiser decisions, and it offered a welcome boost to the adrenaline for the officer confined to desk work. Kenney flew over the Owen Stanleys to look things over shortly after he reached the theater, and MacArthur "bawled hell" out of him for it.[14] Kenney in turn ordered his senior officers to stay out of combat. Wilson, Whitehead, and Walker, each in his own time, defied the order.

During his early months in the theater, Walker often flew with his men; but in October, Kenney specifically ordered him to fly no more missions.[15] The issue came up again, in mid-December, when Kenney again ordered both Whitehead and Walker to stop flying in combat. The night before this confrontation, Walker had gone on a reconnaissance mission. Flying under low clouds in the dark, looking for Japanese barges along the coast, the pilot had hooked a wing on a tree and returned to base with three feet missing from the plane's wingtip. Walker defended his action on the grounds that he should go along once in a while to see how his crews were doing. Kenney insisted Walker was excess baggage on a flight but the best bombardment commander he had. In addition, Kenney pointed out, the Japanese had a reputation for brutal torture of prisoners to obtain information. "I told Walker that frankly we didn't believe he could take it without telling everything he knew," Kenney wrote, "so I am not going to let him go on any more combat missions."[16]

Walker could not have relished having his personal endurance thus questioned by his commanding general, regardless of Kenney's motives for protecting him. Did he choose to go on the 5 January mission to prove himself? Or was he merely following his personal inclination to take risks? He impressed those who knew him best as an intense individual—driven, even obsessed. Airpower's pioneers were not, as a rule, tempered people. They were crusaders, fanatics, believers, visionaries, unwilling to accept the status quo or the limits of the known. Fervent belief exacts a price, and we can see it taking its toll in Walker's life. Like many high-strung individuals, he thrived on challenge, whether physical or intellectual. According to a reporter's account, this would be the 17th mission he had flown since coming to the theater.[17]

Was he acting out some fatalistic wish to end his life? At their farewell lunch at the Willard Hotel in Washington, he had suggested to Bob Pearson that he was going to war with little expectation of returning. "I doubt if I'll be back," he had said. "I've made a terrible mess of things here."[18] Mamie Lee Andrews believed that he had come to regret his divorce from Marguerite; she felt certain that had he come back from the war, they would have gotten back together.[19] Marguerite herself indicated, many years later, that Walker thought he had messed up his life. "He felt he'd made a mistake, you know." He might have expressed this to her when talking in the kitchen on that last evening before leaving the States, but it is also possible that Marguerite elected to believe this because it assuaged her own regret. In any event, her interpretation of this deliberate unnecessary risk of his life was, "I think he didn't care."[20]

On the other hand, Mary Hodge, who knew him less long but saw a great deal of him in 1941, did not perceive that Kenneth Walker felt any sense of failure. To the contrary, she concluded that he was not only quite happy in his new personal relationship, but that he enjoyed a great sense of satisfaction and worth derived from his career and his successful role within it. He had talked with enthusiasm of writing about his work after the war, despite the joking disclaimer, "If I come back." She could recall no evidence that he dwelt upon the idea of not returning. "Almost every man who went to war," she said, "felt that he wouldn't come back."[21]

Whatever else entered into Walker's decision to accompany the 5 January mission, we can be certain that he wanted to make firsthand observations of a large daylight raid. One of the crew, Donald L. Sanxter, heard him tell the crew before takeoff that he had a new camera and hoped to set some good pictures.[22] More than likely, Walker believed, as all men going into combat need to believe, "it won't happen to me." He had compelling cause to go with his men. He was in the middle of the first practical test for the ideas that had propelled his entire career. He wanted to see it through, to see his views vindicated, to see airpower come into its own. Throughout his life, he had subjugated his personal goals to his professional goals; the latter must therefore be attained for his life to have meaning. To hold back now would be untrue to the man he had become.

It may have been for those same reasons that Walker ordered the raid for noon in direct defiance of Kenney's orders. Or was it an act of supreme arrogance, an extreme expression of the courage of conviction? Stubborn, Kenney had called him. Rabid, the officers at the ACTS had called him. "If it was clear in his own mind that was the way to do it," Mary Hodge said, "he would have done it that way." There is evidence, however, that his motivations were appreciably higher. One of his considerations was that most of the crews preferred a total daylight mission.[23] Throughout the preceding months, he had shared their dangers and provided personal, participatory leadership. Now, as always, he sought to balance their welfare with the task to be done. Bad weather, however, which definitely affected Walker's final decision, may well have been the deciding factor. As it turned out, the rain at Iron Range was so heavy the B-24s there could not take off at all. Those from Moresby—six B-17s, six B-24s—went without them.[24]

One of the briefing officers for the mission was Maj David Hassemer, who had often flown on Walker's wing when the two were stationed at Wheeler Field. It was Hassemer who had inherited Walker's "Gold Bug" after losing his own plane. Before the briefing, the two men had a few minutes to reminisce about those days and catch up on mutual acquaintances. Walker told Hassemer he was looking forward to the mission, not having had many opportunities to fly combat. That morning, he was listed for the lead plane: B-17 #41-24453 from the 64th Squadron, 43d Bombardment Group, piloted by Lt Col Jack W. Bleasdale, the 43d Group's executive officer. Other officers aboard were Maj Allen Lindberg, 64th Squadron commander, and Capt Benton H. Daniel. After looking at the loading lists for the aircraft, Hassemer objected to Walker being in the same plane as the deputy group commander and a squadron commanding officer. His concerns were dismissed and he was overruled, much to his later regret.[25]

The B-17s took off about 0800 from Jackson Field, one of the three main fields in the Moresby area. They rendezvoused with the B-24s and led the way toward Rabaul with the

Liberators visible behind them. They encountered some "light stuff" over the Owen Stanleys, but otherwise the weather was clear.[26]

They dropped their bombs on Rabaul from approximately 9,000 feet at noon. The Japanese "were caught unawares, more or less," but the antiaircraft fire was "moderate to heavy and ranged to accurate."[27] Each plane bombed individually; each crew selected its own target.[28] The B-17s had completed their run by the time the B-24s came in. Walker's plane, the lead B-17, was hit and had dropped out of formation before the B-24s were over the target.

When his B-24 came in to attack, navigator William Whitacre "saw at least 3 ships hit in the harbor & left burning. Heavy ack-ack & Zero interception which lasted half an hour. We went in over the harbor to bomb ships & could see the airdrome with Zeros taking off to come up & get us & bombers taking off to keep from being bombed." As his plane went into its bomb run, Whitacre saw a B-17 "losing altitude & at least one engine was smoking."[29]

Sanxter also saw a B-17 circling below them with several Japanese fighters attacking it. He assumed Walker was in the circling plane, trying to take unobstructed pictures of the damage and "get some good evidence of the result of daytime bombing to show General Kenney." None of the other planes went to the B-17's assistance, a fact Sanxter explained in terms of the pilots' mind-set. Because most of their missions had been by single planes and/or at night, the pilots "tended to think more as single-plane units rather than elements," he wrote. Against enemy fighters, their experience had been that they "could handle [themselves] and escape destruction" in part because the Japanese fighters used a small caliber bullet that inflicted fatal damage only if it hit a vital part. Consequently, the pilots tended to concentrate on avoiding the antiaircraft fire, which was potentially more dangerous to them. Under the circumstances, Sanxter believed, others would have assumed that the B-17 pilot "could handle the situation and would rejoin the group if he thought it necessary." In retrospect, Sanxter believed the B-17 was observing the proceedings and underestimated the severity of the attacking force.[30] The plane did not rejoin the formation. It was last seen headed south, just east of Vunakanau airdrome, at an altitude of approximately 5,000 feet. The left outboard engine was smoking and out.[31]

The 5 January raid was successful in terms of damage inflicted. The crews reported hitting 10 vessels and probably sinking one. The Japanese later acknowledged losing the 5,833-ton *Keifuku Maru* on that date.[32] Two of the Fifth's bombers did not return. One was the B-17 in which Walker had flown. The other was a B-24 from the 321st Squadron piloted by Walter Higgins, who went down on his return flight and radioed in to give his approximate position.

Kenney's diary entry was terse. "Walker off late. Disobeyed orders by going along as well as not starting his mission when I told him." He ordered the Catalinas and other reconnaissance planes to search the Trobriand Islands area and the route to Rabaul. Returning crews had reported the plane as last seen "headed south near Wide Bay with an engine oil fire and 2 Nips on its tail." Shortly after the search planes went out, a report came in that a plane had been spotted on a coral reef in the Trobriands. Anticipating that his bomber commander would shortly be rescued, Kenney vented his anger by telling MacArthur that he was going to reprimand Walker officially and send him to Australia on leave for a couple of weeks.

"If he doesn't come back," MacArthur said, "I'll put him in for a Medal of Honor."[33]

On his return to Moresby from the 5 January raid, James A. McMurria, 321st Squadron, set out again on the search. When he spotted a downed plane and its survivors, he circled and dropped a note, hastily printed in large letters:

"Milne & Moresby notified - - - Expect flying boat within 24 hours. This is McMurria & crew of 321st. If you are Higgins all go to end of Islet & face the large Islands on the South - if General Walker all go to North End of Islet - If Another crew all stand in circle." The men were Higgins and crew, less two who died in the crash. They were picked up by a Catalina the next morning.[34]

On 6 January, two B-17s, one B-24, and five B-26s that were searching for Walker were attacked, although not hit, by Japanese fighters. When Allied reconnaissance spotted a

Japanese convoy headed toward Lae from Gasmata, no Allied forces were available to attack it. "Most of our B-17s out hunting for Walker," Kenney wrote. "Could not get them back in time for strike on the convoy today." The problem of limited resources was inescapable. "We will keep on searching," Kenney wrote.

Three weeks later, Kenney wrote Kenneth Walker Jr., urging him not to give up hope for his father. "There is still a chance that he may have been taken prisoner."[35] In his private diary, however, Kenney was more grim. "I have no hope for Ken" he wrote on 6 January. "That ship of his was shot down in flames and unless the crew bailed out they are gone. If they bailed out near the shore the Japs have them and that is bad."[36]

Kenney canceled a scheduled trip to Bena Bena, where the natives had planned a big "sing-sing" to honor the "Number One Baloose Man" ("baloose" was pidgin for airplane). Instead, he concentrated on the Walker search plus attacks on the Lae convoy.[37] The search yielded nothing, but attacks on the convoy were successful; and the ships that made it to Lae were pounded mercilessly after arriving. On 9 January, MacArthur issued a communiqué praising the forces under his command for the victory that had been achieved at Buna. He announced Distinguished Service Crosses for 12 officers, including Walker, who had made it possible.

The men of Fifth Bomber Command reacted soberly to Walker's loss. John Perakos had admired him for his active role in combat, which he thought raised him in the eyes of his command. Now that he was missing, Perakos began to think his expertise might have been better used in planning. Patrick Norton was "shocked" to hear that Walker was over Rabaul in the first place, because "we didn't have rank to spare." By the evening of 9 January, few entertained hopes that Walker might still be found alive. William Travis, who followed news of the unsuccessful search with personal regret, wrote in his diary that night, "General Walker is presumably lost."

Due to the combination of rough country, shark-infested waters, and limited organization and/or means for rescue, the recovery of any downed crew was problematical. The understanding among the airmen was that a bailout over land was a last resort. Few men were ever recovered from the mountainous jungle, where "all you had to help you was your feet." "If somebody got down in the jungle and we could find him," remembered Norton, "we'd break our backs to get him out." On the other hand, he admitted, "We couldn't do all that much." Bailout near enemy territory was equated with a slow death in a Japanese POW camp. The men rode their planes down and ditched in the ocean when possible, hoping to survive on life rafts until they could be found.[38]

For Walker, time was running out. On 11 January, MacArthur's headquarters notified the family and released the news that he was missing. Walker's record "has been outstanding," MacArthur stated, "and much of the efficiency of the bombardment command of the 5th American Air Force is due to his exceptional brilliance and courage."[39]

Fifth Bomber Command was placed temporarily under Whitehead, whom Travis characterized as a "one-man offensive" whose "idea of a fight is to throw every available aircraft against any target." Kenney was dissatisfied with his bomber command. He put in a request for Brig Gen Howard Ramey. He believed Ramey had "a good steady hand" and would straighten out the Fifth's troubles.[40] But Ramey was to have too little time to see what he could do. On 25 March 1943, he and his crew "disappeared without trace" on a reconnaissance flight over New Guinea.

"It appears that we are particularly vulnerable so far as commanders for bomber command are concerned," Wilson wrote Whitehead. "As in the case of Ken Walker, I am still hoping that these people will turn up on the beach someplace."[41]

Walker did not turn up on a beach. It had been his lot to serve Fifth Bomber Command during its days of greatest crisis and fewest resources. Not long after he disappeared, the Southwest Pacific's air forces used skip bombing and low-altitude attacks to score a significant victory in the Battle of the Bismarck Sea—a turning point that marked the end of their darkest days.

Walker's Last Photograph

WESTERN UNION

SAC124 50 GOVT=WASHINGTON DC 10 847P

KENNETH N WALKER=

1009 EAST WINDSOR ST GD=

THE SECRETARY OF WAR DESIRES ME TO EXPRESS HIS DEEP REGRET THAT THE COMMANDING GENERAL UNITEDSTATES ARMY FORCES IN THE SOUTHWEST PACIFIC AREA HAS REPORTED YOUR FATHER BRIGADIER GENERAL KENNETH N WALKER US ARMY AS MISSING IN ACTION SINCE JANUARY 5 PERIOD ADDITIONAL INFORMATION WILL BE SENT YOU WHEN RECEIVED=

ULIO THE ADJUTANT GENERAL.

5 ULIO. THE COMPANY WILL APPRECIATE SUGGESTIONS FROM ITS PATRONS CONCERNING ITS SERVICE

Missing in Action

Notes

1. Wesley Frank Craven and James Lea Cate, eds., *The Army Air Forces in World War II*, vol. 4, *The Pacific: Guadalcanal to Saipan August 1942 to July 1944* (1950; new imprint, Chicago: University of Chicago Press, 1948–1958), 135–138.

2. Kenney Papers, vol. 4, Office of Air Force History (OAFH), 1 January 1943..

3. Historical Records Survey (Inventory) 222, 223, 251, 226, National Archives; and Kenney Papers, OAFH, 1 January 1943.

4. Assistant Chief of Staff, Intelligence, *Air Action in the Papuan Campaign, 21 July 1942 to 23 January 1943* (Maxwell AFB, Ala.: Air University Library, Documents Section, M-U27218 no. 17, 1944), 64–65.

5. Kenney Papers, OAFH, January 1943; and Diary of William L. Whitacre, 4 January 1943.

6. Kenney Papers, OAFH, 4 January 1943.

7. Patrick Norton, telephone interview with author, August 1991.

8. Craven and Cate, vol. 4, 111 n.

9. Kenney Papers, OAFH, 10 December 1942.

10. John Perakos, telephone interview with author, 12 August 1991.

11. Donald Wilson, oral history interview with Hugh N. Ahmann, K239.0512-878 and K239.0512-810, USAFHRC, 212–14.

12. Lex McAulay, *Battle of the Bismarck Sea* (New York: St. Martin's Press, 1991), 1–4.

13. Kenney to Arnold, letter, 1 January 1943, 706.168-2, USAFHRC.

14. Kenney Papers, OAFH, 21 August 1942.

15. Kenney Papers, OAFH, 5 October 1942.

16. George C. Kenney, *General Kenney Reports: A Personal History of the Pacific War, With Maps By the Author* (1949; new imprint, Washington, D.C.: Office of Air Force History, 1987), 166–67.

17. Haywood S. Hansell Jr., "Brig. Gen. Kenneth N. Walker, Prophet of Strategic Air Power," *Air Force Magazine*, November 1978, 94; and "With his Boots On," *Time*, 41, 5 April 1943, 73.

18. Scott Pearson, telephone interview with author, 7 February 1991.

19. Mamie Lee Andrews, telephone interview with author.

20. Marguerite Walker, videotape interview.

21. Mary Hodge, interview with author, 16 October 1990.

22. Donald L. Sanxter to Gene Monihan, letter, 24 February 1991.

23. Ibid.

24. Diary of William Whitacre, 5 January 1943.

25. David W. Hassemer, interview with author, 21 January 1991.

26. Sanxter to Monihan.

27. Whitacre diary; and Sanxter to Monihan.

28. Sanxter to Monihan.

29. Whitacre diary.

30. Sanxter to Monihan.

31. Adjutant General (AG) 704-Missing, SR&D 299, RG 407 National Archives (NA).

32. Craven and Cate, vol. 4, no. 25, 716.

33. Kenney Papers, 0AFH, 5 January 1943.

34. James A. McMurria to Gene Monihan, letter, 30 May 1991.

35. Kenney to K.N. Walker Jr., letter, 28 January 1943.

36. Kenney Papers, OAFH, 6 January 1943.

37. Kenney Papers, OAFH, 3 and 6 January 1943.

38. John Perakos, telephone interview with author, August 1991; Patrick Norton, telephone interview with author, August 1991; Diary of William Travis, 5–9 January 1943; William Travis telephone interview with author, 4 August 1991; McAulay, 15–16; and Sanxter to Monihan.

39. Associated Press dispatch, 11 January 1943.

40. Diary of William Travis, 7 January 1943; Kenney, 131; and Kenney Papers, OAFH, 18 January 1943.

41. Enis Whitehead to Donald Wilson, letter, 27 March 1943; Wilson to Whitehead, letter, 27 March 1943; Whitehead personal letters, 168.6008-3 USAFHRC.

Chapter 7

The Lingering Doubts

MacArthur recommended that Walker be awarded the Medal of Honor for "conspicuous leadership above and beyond the call of duty, involving personal valor and intrepidity at an extreme hazard to life." MacArthur's citation noted that Walker repeatedly accompanied his units on bombing missions deep into enemy territory. "From the lessons personally gained," it went on, "he developed a highly efficient technique for bombing when opposed by enemy fighter airplanes and by antiaircraft fire."[1]

The Air Force was such a young branch of the service that its customs and expectations were not yet fixed. Upon receipt of MacArthur's recommendation, the Adjutant General addressed a query to Arnold: "Is it considered above and beyond the call of duty for the commanding officer of a Bomber Command to accompany it on bombing missions against enemy held territory?" Although his papers do not reveal what discussions, if any, led to the decision, the reply came from Maj Gen George Stratemeyer, chief of the air staff, on 29 January. "It is the considered opinion of Headquarters, Army Air Forces," he wrote, "that the conspicuous leadership exemplified by Brig Gen Kenneth N. Walker on the specific mission as cited by General MacArthur does constitute action above and beyond the call of duty."[2] The recommendation was approved as written in MacArthur's headquarters. Walker became World War II's 43d recipient of the Congressional Medal of Honor. Only 38 Medals of Honor were awarded to flying personnel of Army Air Forces during the war.[3]

Kenneth Walker Jr., 16 years old, received the Medal of Honor from President Roosevelt in the Oval Office on 25 March 1943. Later in the year, 10-year-old Douglas received, in his father's name, the Legion of Merit, awarded to Walker in recognition of his work in the Plans Division between July 1941 and March 1942.

The awards were not made posthumously because Walker was listed as missing, not dead. In March, Walker's family received a letter saying Tokyo radio had announced that Walker and his entire crew had been taken prisoner.[4] Thus

THE WHITE HOUSE
WASHINGTON

CITATION FOR MEDAL OF HONOR

KENNETH N. WALKER, BRIGADIER GENERAL, UNITED STATES ARMY. FOR CONSPICUOUS LEADERSHIP ABOVE AND BEYOND THE CALL OF DUTY INVOLVING PERSONAL VALOR AND INTREPIDITY AT AN EXTREME HAZARD TO LIFE. AS COMMANDER OF THE FIFTH BOMBER COMMAND DURING THE PERIOD FROM SEPTEMBER 5, 1942 TO JANUARY 5, 1943, BRIGADIER GENERAL WALKER REPEATEDLY ACCOMPANIED HIS UNITS ON BOMBING MISSIONS DEEP INTO ENEMY HELD TERRITORY. FROM THE LESSONS PERSONALLY GAINED UNDER COMBAT CONDITIONS HE DEVELOPED A HIGHLY EFFICIENT TECHNIQUE FOR BOMBING WHEN OPPOSED BY ENEMY FIGHTER AIRPLANES AND BY ANTIAIRCRAFT FIRE. ON JANUARY 5, 1943, IN THE FACE OF EXTREMELY HEAVY ANTIAIRCRAFT FIRE AND DETERMINED OPPOSITION BY ENEMY FIGHTERS, HE LED AN EFFECTIVE DAYLIGHT BOMBING ATTACK AGAINST SHIPPING IN THE HARBOR AT RABAUL, NEW BRITAIN, WHICH RESULTED IN DIRECT HITS ON NINE ENEMY VESSELS. DURING THIS ACTION HIS AIRPLANE WAS DISABLED AND FORCED DOWN BY THE ATTACK OF AN OVERWHELMING NUMBER OF ENEMY FIGHTERS.

Franklin D Roosevelt

Medal of Honor Citation

began an uncertainty that has endured for half a century. Walker's plane has not been located, nor has indisputable evidence been presented either that he died in the crash or survived to be taken prisoner. The record is sparse.

According to War Department files, the father of Major Lindberg, who was lost in the same plane as Walker, talked with the pilot of the plane that flew on the missing plane's right wing. Based on his perception of the plane's condition when he last saw it, this pilot gave his opinion that the plane might have made a safe landing in enemy territory. He also told Lindberg that they picked up a Japanese radio broadcast

Ken Walker Jr. Receives His Father's Medal of Honor

about three days after the mission, to the effect that Walker had been captured. Since Kenney did not announce until 11 January that Walker was missing, this suggested to some that the crew had indeed been captured. Reports that the crew might be prisoners were noted at the 64th Squadron headquarters, but their veracity was not checked. Upon reviewing the relevant facts in January 1944, one year after the men disappeared, the Status Review and Determination

Doug Walker Receives His Father's Legion of Merit

Section of the Adjutant General's Office (SR&D) concluded, "it seems highly possible that the descent of the plane to land on water could have been observed by the enemy and in that event, persons escaping by parachute or surviving the forced or crash landing in any manner, would have been captured." The report noted that other crew members were also unofficially reported to be prisoners. These men "may be reasonably presumed to be living." They were therefore continued on the list of those missing in action.[5]

After Japan surrendered in August 1945, the Army took immediate steps to locate prisoners and recover remains of those deceased. A Missing Personnel Investigation Unit (MPIU) landed at Rabaul on 10 September 1945, along with the first elements of the occupation force. Despite making missing personnel its top priority, MPIU found nothing. Most chilling, they found some clear evidence that natives had at times betrayed prisoners to the Japanese and no evidence that natives would have helped men who were downed. Japanese records had been destroyed, and no complete list of prisoners could be obtained. While planning to continue searches and inquiries, MPIU nevertheless concluded that "action for death presumption for the men reported missing over New Britain must inevitably be taken."[6]

In December 1945, the SR&D reviewed the case of Walker and his 10 fellow flyers. New information had surfaced. An International News Service war correspondent, Lee Van Atta, went to New Britain following Japan's surrender and interviewed Allied prisoners of war who had been held at Rabaul. He filed a lengthy dispatch that concentrated primarily on the hardships the men had endured. Since he used the editorial "we" rather than direct quotations, the source for a key paragraph could not be verified. The paragraph read, "We knew the details of the death of Brig. Gen. Kenneth Walker, commander of the Fifth Bomber Command, too. We learned them from one of the surviving crew members of his plane."

The American flyers identified in the news story were James A. McMurria, Jose L. Holguin, and Alphonse D. Quinones. When those three men returned to the States, they were asked to tell what they could about the Walker case. McMurria, who had been on the 5 January raid and later piloted the B-24 that located Higgins and crew, was himself shot down at sea (on 20 January), captured, and taken to Wewak, New Guinea, for interrogation. "I was shown documents signed by Major Bleasdale which establishes the fact that Major Bleasdale, at least, survived the crash." When the Japanese asked McMurria if he knew of General Walker's death, McMurria figured they had gotten the information that he was dead from Bleasdale. McMurria, who was later taken to prison at Rabaul, never saw Bleasdale. Responses to his inquiries about Walker while at Rabaul were all negative.[7]

Holguin reported that he was shot down in June 1943 and "[I] was finally captured on July 17, 1943, and the Japs attempted to interrogate me." They wanted the name of the commander of Fifth Bomber Command. When Holguin refused to give them the information, "they presented me with a book containing some information on Major Bleasdale and General Walker. They said it was no use trying to lie; that they knew General Walker was dead and Major Bleasdale captured. Therefore, they wished to know my new commander's name. From what I could gather,

General Walker was killed while still in the air, after his plane was hit by Jap zero fighters. The plane caught fire and Major Bleasdale managed to get out somehow. General Walker, being dead aboard the ship, had to go down with it."[8]

Quinones became a prisoner in December 1943. He was questioned about the commanders at Moresby several times during his captivity. "In several instances, I was asked if I knew or had heard of General Walker. It seemed to me the Japs had no definite information about his death or rescue and were seeking confirmation one way or the other." His own opinion, based on what he heard from others, was that Walker was dead.[9]

SR&D noted that none of the three had talked with a survivor of the 5 January crash, as Van Atta's news story had suggested; nor did they have any information about Walker's death except what the Japanese gave them. SR&D queried the Manila headquarters to see if any additional information might be gleaned, but the reply was based on the Van Atta account and offered nothing new. The board concluded that the crew of 11 could not "reasonably be presumed to be living" and recommended that findings of death be made. This was done, bearing the official date of 12 December 1945.[10]

In May 1948, the secretary of the Army, through the adjutant general, requested that the Philippine-Ryukyus Command make an effort to locate the wreckage of the plane and ascertain the fate of each crew member. During December 1948 and January 1949, the 604th Quartermaster Graves Registration Company conducted an extensive search and investigation for the several hundred aircraft still missing in New Britain and surrounding waters. The only information relevant to Walker's case came from Father Poncelet, who—along with a number of nuns from the Order of Our Lady of the Sacred Heart—had been captured in the Japanese advance and held prisoner at Rabaul. Father Poncelet stated that Captain Daniel, a member of Walker's crew, had been brought into his POW camp on 12 January 1943. The quartermaster company recovered a number of remains; but even though they searched all the villages surrounding Wide Bay, where Walker's plane was deemed most likely to have gone down, it was not found.

In April 1949, the files were reviewed and a memorandum for the record prepared. It noted that the legal section from the Philippines Command had closed the case and planned no further action. After a careful review of the little that was known, the investigator concluded the information did not warrant issuing an official report of death. When a Board of Review brought up the Walker file on 20 July 1949, it resolved that Walker and the other 10 be declared nonrecoverable. For the War Department, the case was closed.[11]

The case also seemed closed for his sons. Late in 1943, the War Department sent Harold L. George, Walker's executor, his checkbook and four manila envelopes of personal papers. George said he would send them on to Kenneth after examination, but Kenneth does not remember receiving them and efforts to locate them have been unsuccessful. The Army sent Kenneth his father's personal effects—his clothes, a cigarette case, an all-silver ash tray, a bronze baby shoe, a box of photographs, a phonograph record, and a songbook. Both he and Douglas declined the opportunity to be appointed to West Point on the grounds of their father's Medal of Honor. Their half-brother John, whom they hardly knew, took the appointment and turned the opportunity into a career in the US Navy. Kenneth became a college professor, Douglas an expert in communications.

As an adult, Douglas became interested in his father's career. He found little in government files, however, and visits with Mary Hodge confirmed that Walker's personal papers—his entire professional life, as she had described them—had been destroyed by the tenants who rented the Hodge house in the years immediately following the war.[12]

Douglas also considered retaining an artist to paint a scene depicting his father's last mission. A friend, Gene Monihan, began gathering details of the mission so as to have it portrayed as accurately as possible. During this process, he talked with Mary Hodge's son-in-law, Charles Barton, who recalled overhearing a conversation that took place in Japan between Mary Hodge and Sir Mark Young, the former British Governor General of Hong Kong who had spent the war years in Japanese prisoner of war camps. Young asked about Walker, saying he had been in a prison camp near Shanghai with him. As Barton recalled the conversation, Mrs. Hodge told Young that Walker had been killed early in the war, and the conversation moved to other topics.

Mary Hodge, questioned about the conversation some 30 years later, did not recall it. Sir Mark Young died in 1974, and an inquiry to his estate by Douglas Walker yielded no information. Douglas queried Jose Holguin, who reaffirmed that during his interrogation the Japanese told him Walker had been killed. "I was given to understand that they had found his body in the wreckage of his airplane. Those and other crew members were then buried nearby." Based on his experience and knowledge of Japanese procedures, Holguin believed the Japanese account to be "probably true." He never heard any of the other prisoners speak of Walker as a POW. Nevertheless, as so often happens when a body is not recovered and there is no tangible evidence of death, the question seemed unresolved. Did Walker survive the crash?[13]

Douglas Walker and Gene Monihan queried Bruce Hoy, curator of the Aviation, Maritime, and War Branch of the National Museum and Art Gallery in Papua, New Guinea. Hoy had done extensive research and recovery work on the Allied planes that had been downed in the area. He had also assisted the US Army's Central Identification Laboratory (CIL), a highly specialized unit whose primary mission was the recovery and identification of the remains of those Americans presumed killed during the Vietnam War. CIL began investigating World War II losses in the Pacific in the early 1980s, when poor relations between Vietnam and the United States left it with little to do in Vietnam.[14] Hoy was not optimistic about the possibilities of finding the Walker plane. He noted that no exact crash location had been pinpointed and that planes often flew for some distance after being disabled. The area where he deemed the plane most likely to be is wild and uninhabited. He had no specific leads.[15]

As Hoy, Monihan, and Douglas Walker exchanged information relevant to Kenneth Walker's case, they found some translations of Japanese interrogations of prisoners. Intelligence Record No. 14, dated 26 February 1943, concerns interrogations of two men who appear to be Bleasdale and Daniel from the 5 January raid. The narrative summary at the beginning of the first interrogation indicated that the prisoner (Bleasdale ?) was in a B-17 which attacked Rabaul. Japanese fighter planes damaged the left engine; the plane lost altitude and circled southward. "Realizing his danger," the translator's narrative reads, "PW took to his parachute over a mountain north of Wide Bay. Later, while wandering through the mountains, he discovered a native hut where he hid and rested for 20 days. The natives said there was a British missionary at Wide Bay. On the way there, PW was captured through efforts of Navy observation post personnel at Zungen, 28 January." If three days are allowed for the early wandering in the mountains plus their later journey to find the missionary, the time frame fits the Walker raid.

The translator went on to record that the prisoner was sick and was interrogated on only a few matters. The interrogation reveals that he was a pilot and that "Capt X, who was captured with this PW, was sent to Kavieng by the Navy."

The second interrogation on 26 February, presumably of Capt X, contained background information identical to the first, indicating that the two men had been together. The captain was queried extensively about the organization, command, and strength of Fifth Air Force. Under Fifth Bomber Command, Maj Gen Walker is listed as the commanding officer. Beside this, in parentheses that presumably indicate a translator note, is written, "According to B information, he was shot down." The "B" could refer to Bleasdale. Daniel was the only captain in the fated crew.

That the two prisoners in this interrogation were Bleasdale and Daniel was also the conclusion of Lex McAulay, Australian author of *Battle of the Bismarck Sea,* who sent Hoy details about the prisoners he had talked with.[16] The fate of neither Bleasdale nor Daniel is known; neither returned from Japanese POW camps after the war. The interrogation, however, provides one additional bit of evidence that Walker was killed in the crash. (Monihan believes it possible that the pilot who was interrogated was Walker himself, and that he tried to hide his identity but the Japanese learned who he was and transferred him to POW camps in either China or Japan, where he probably died during captivity.)[17]

It seems unlikely that the exact circumstances of Kenneth Walker's death will ever be known. In any event, they do not alter the substance of his life. He earned his place in airpower history by the depth of his belief and commitment. Having concluded that the bomber would always get through, he set his sights on the specifics of the bombing mission. He pursued his course with a persistence and perseverance that gave heart to those less certain. He neither wavered nor deviated. He led his fellow officers in formulating a way whereby a new, unexplored force could be translated into a means of national policy. It seems fitting that his name is inscribed on Walker Hall, the College of Aerospace Doctrine, Research, and Education at Air University in Montgomery, Alabama.

College of Aerospace Doctrine, Research, and Education

Notes

1. 13 January 1943; copy in K141.2421, United States Air Force Historical Research Center (USAFHRC).

2. Adjutant General (AG) to CGAAF, 14 January 1943; and reply 29 January, K141.2421, USAFHRC.

3. Philip S. Gage Jr., "Fliers and Medals of Honor," *Aerospace Historian* 30, no. 4 (December 1983): 254–60. Of the total 3,419 awarded up to 1 January 1980, Gage shows 105 awarded to flyers.

4. Transcript, Bill Henry's "By the Way" broadcast, 25 March 1943. Newspaper clippings.

5. AG-704 Missing Report, 5 January 1944, National Archives (NA).

6. Reports of Missing Personnel Investigation Unit, 706.288, USAFHRC, 57.

7. AG-704 Missing Report, 12 December 1945, NA; and James A. McMurria to Gene Monihan, letter, 30 May 1991.

8. AG-704 Missing Report, 12 December 1945, NA.

9. Ibid.

10. Ibid.

11. AG-704 Missing Report, Board Proceedings 2352, 20 July 1949, NA. Also Personnel Command File, Memorandum for Record, 29 April 1949.

12. Conversations with Mary Hodge.

13. Gene Monihan, telephone interviews with author, 8 August and 15 October 1990; Monihan to McMurria, letter, 11 August 1991; Charles L. Hodge interview with author and diary of Charles L. Hodge, 16 October 1990; conversations with Douglas Walker; and Jose L. Holguin to Walker, letter, 19 September 1982.

14. Susan Sheehan, "A Missing Plane," *The New Yorker* 62 (12 May 1986): 46-48; 62 (19 May 1986): 43-44; and 62 (26 May 1986): 39-40.

15. Bruce Hoy to Gene Monihan, letter, 23 May 1988.

16. Hoy to Monihan, letter, 3 June 1988.

17. Monihan to Hoy, letter, 24 June 1988.

Appendix A

Kenneth N. WALKER

MEMORABILIA

March 1st.

Dear Ken -

Thanks for the letter. You certainly are a busy one. The pictures were appreciated. Indeed! Pipe the curly hair. The Hawaiian touch, perhaps.

Roger Ramey's wife is in Washington. Saw her the other evening. She with many other Army wifes were recently evacuated from Hawaii. Told some interesting sidelights on the attack there. She said she was awakened by a booming noise and thought it the Coast Artillery practicing. Got up and looked west out of her window and saw two planes flying low and right in front of the window. One was on fire. One plane turned and she saw the rising sun on the side and knew the Jap attack must be on.

At Wheeler Field one young 2d Lieut. rushed down to the line with his pal who was going to try and take-off. The latter got into a plane but when taxiing out a Jap plane dove on him and killed him with machine guns. So the first lad got into another plane, took off and knocked off two Japs. The funny part of it was that this lad had never been in a pursuit plane before in his life - had some difficulty in getting checked off on the AT-6's. Did pretty well for his first flight.

A grand new War Dept. reorganization is to take effect on the 9th. It doesn't please us very much. Particularly myself - as I shall have to go up to the new superior general staff - in the War Plans Division. I'm pretty unhappy over it. There will only be 40 officers on the W.P. Division of which 20 will be air officers. I shall try very hard to remember your "pep talk" about "they also serve who only sit and wait" or something like that, and try and get some comfort out of the thought.

Some of the officers whom you met are now out in the Java area. Have been doing fine work. A whole raft of decorations was awarded to them the other day and they have well deserved them.

Keep up the good work and get the eagle rating. And keep the old nose to the grindstone to the extent of getting the school work done. How about the athletics. Suggest that you take up tennis, Ken. Its something that one can do all his life. I went in for boxing and of course I couldn't keep that up. I'd like both Doug and you to box and be good at it. Tennis should help your eyes and you can play with glasses.

Tell Doug that I certainly appreciated his letter and to write often.

Love to my boys,

Dad

Letter from Dad

WAR DEPARTMENT
WAR DEPARTMENT GENERAL STAFF
OPERATIONS DIVISION
WASHINGTON

May 1 1942.

Dear Ken -

Thanks for the letter. Glad to note that both of my boys are doing all right in school. Tell Doug that being tardy is like being late for a formation - and that guys in my command who are late at formations have unhappy incidents in their lives. But I'm sure that there was a good reason for it. The progress report was fine - particularly the last two entries which are important in life.

Your grades are OK too. Getting good grades do not indicate necessarily what one learns, however. So the real test is the test one gives himself - "am I getting the most out of my opportunities".

I confess that I dont know what the light on the P-38 was. I'm getting to be quite a Kiwi - for I dont get out amoung combat outfits and dont often see our later planes.

The Operations Division is the new name for the War Plans Division of the War Depart ent General Staff. It has two main "groups": one the Strategic and Planning Group; the other the Theatre Group. The latter functions as the Chief of Staff's command post for handling the several Theatresof Operations. Gen. Streett (Billy's Dad) is the Chief of the Theatre Group. We try and take care of the commanders in the Field, and all communications from them and to them are handled for the most art by us.

Still expecting to get active service but am afraid I may have to wait for a number of months. A couple of generals are trying to get me pried loose from the General Staff but no luck so far.

Tell Doug to step along and get to be a Lion Cub. And by the way - how are you doin ?

Love to my fine fellas

Dad

FOR VICTORY BUY UNITED STATES SAVINGS BONDS AND STAMPS

By the way - Harry Overturf got a commission in the AC as a 1st Lt. and is here in Washington. He stayed with me till

Letter from Dad

WAR DEPARTMENT
WAR DEPARTMENT GENERAL STAFF
OPERATIONS DIVISION
WASHINGTON

June 15,1942.

Dear Ken and Doug -

Enclosing a press release which indicates that I'm going to be a field soldier after all. Of course the Senate has to confirm the nomination and they may balk when they get to my name. I understand that it should take around five days for the nomination to be confirmed. I'm informed then that I will probably receive orders for over-seas and will have to leave within 48 hours thereafter. Am set up for a Bomber Command. Am told that the previous plans for my assignment have been changed and that now I'm going westward. If so, I shall surely get to see my two fine fellas. I'm counting on that very very much.

Haven't time to write any more. Am trying to "get my house in order" and that is quite a task.

I hope you read of how well our bombers did at Midway. General Tinker was lost on a mission. Lt.Ray Salzarulo was with him - you remember Ray - he lived upstairs in the Apt. next to mine in Honolulu.

Love to my two boys - and let me hear from you

Dad

Letter from Dad

WAR DEPARTMENT
WAR DEPARTMENT GENERAL STAFF
OPERATIONS DIVISION
WASHINGTON

June 21,1942.

Dear Ken and Doug -

Just a note to advise you of my plans. Expect to leave here July 1st unless General Arnold insists that I start sooner. Plans are for July 1st anyway. I plan to go airline to San Francisco via Los Angeles. Would get in LA on the morning of the 2d and would be able to spend the day with you - going on to San Francisco during the evening.

I'm bundling up a lot of my clothes and sending out to you. A lot you cant wear - but some your Mother might have cut down for you. Cloth may be getting scarce these next two years. I'm sending my old Peal boots and woolen riding breeches which I believe will fit you (unless the boots are too small for Ken's big feet).

Please tell your Mother that I am having an allotment effective this month and the Govt will mail the check directly to her.

The Senate failed to discover that I'm a rabid Republican - so confirmed my nomination June 17th.

Be good boys and I'll see you soon and we'll go out and cover the town.

Love

Dad

Thanks for the wire - made me quite proud that my boys think so well of their old man. And the Father's day present. The socks were swell and I know the candy will be.
I'm leaving here at 5:30 PM July 1st and get in LA the morning of July 2d. Will radio you exact time later. Be good guys and I'll see you soon.

Letter from Dad

Walker with Aide

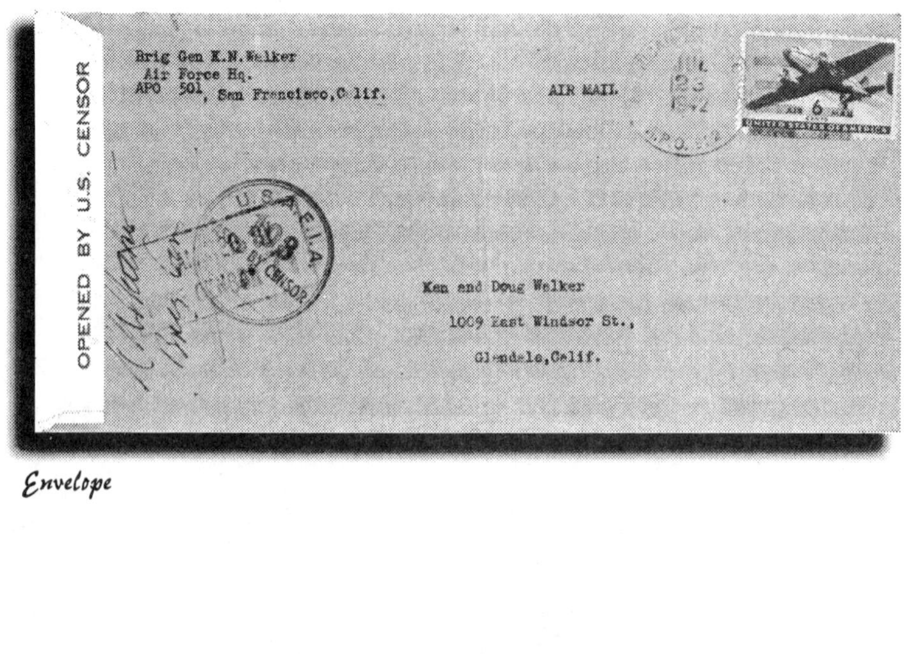

Envelope

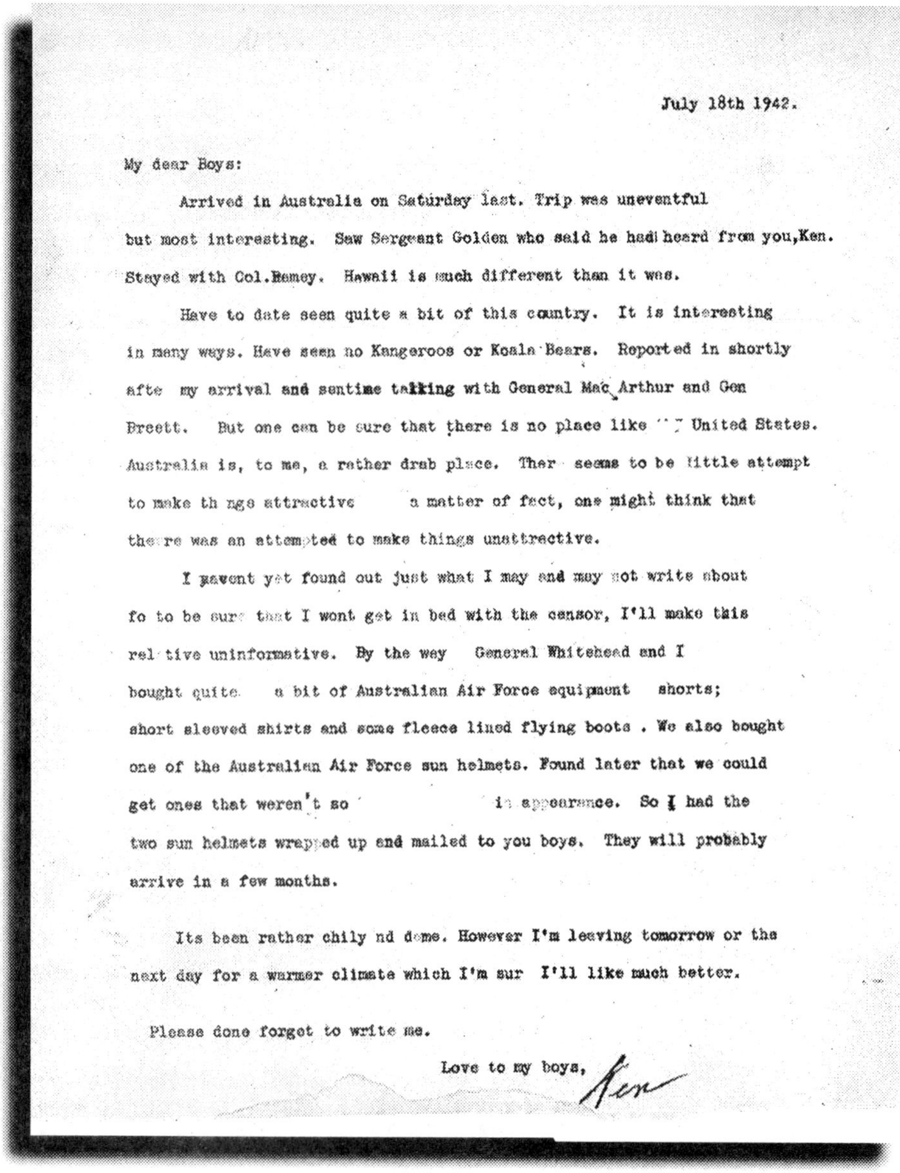

July 18th 1942.

My dear Boys:

Arrived in Australia on Saturday last. Trip was uneventful but most interesting. Saw Sergeant Golden who said he had heard from you, Ken. Stayed with Col. Ramey. Hawaii is much different than it was.

Have to date seen quite a bit of this country. It is interesting in many ways. Have seen no Kangeroos or Koala Bears. Reported in shortly afte my arrival and sentime talking with General Mac Arthur and Gen Breett. But one can be sure that there is no place like [illegible] United States. Australia is, to me, a rather drab place. Ther seems to be little attempt to make th ngs attractive a matter of fact, one might think that there was an attempted to make things unattractive.

I havent yet found out just what I may and may not write about fo to be sur that I wont get in bad with the censor, I'll make this rel tive uninformative. By the way General Whitehead and I bought quite a bit of Australian Air Force equipment shorts; short sleeved shirts and some fleece lined flying boots . We also bought one of the Australian Air Force sun helmets. Found later that we could get ones that weren't so [illegible] in appearence. So I had the two sun helmets wrapped up and mailed to you boys. They will probably arrive in a few months.

Its been rather chily nd deme. However I'm leaving tomorrow or the next day for a warmer climate which I'm sur I'll like much better.

Please done forget to write me.

Love to my boys,
Ken

Letter from Dad

Allied Air Forces,
APO 922 San Francisco Calif
Aug 11 1942.

Dear Boys -

Have been here a month now. Have gotten to see much of the east coast of Australia, the Nowthwest and New Guinea. Dont care much for Australia - things are a little drab. Haven't s een any Kangaroo - however did see one Wallaby(midget kangaroo) when I was over in the Darwin area. Havent been assigned to any specific command yet. A reorganization is in progress. Am informed that will probably get the bomber command.

Have been able to participate in three combat missions with the bombers. Was in a B-17 the first time and we were the only plane participating as we were on a reconnaissance mission endeavoring to locate a convoy. We ran on the convoy while at 6000 ft - circled to get 10000 feet altitude. Fortunately there were no Zeros around (although I was foolishly disappointed for a while - although we had seen a couple about an hour previously) The AA opened up and I got quite a kick out of watching the guns flash and a few seconds later see puffs right below us or to the sides. We got one fair sized hold in the right wing and sometimes the shell fragments sounded like hail on the wings. It was my first time under fire but I was so interested that I forgot to feel concerned. On neither of the other two missions did I see any Zeros. However, before we get to Tokio I'll probably see plenty.

Was up at a Pursuit Group in the NW last week. The Group has shot down 60 jap planes and has lost only five in combat - a pretty good record. All the pilots are young kids - with a fine spirit. One of the pilots I met has 10 japs to his credit - a couple of others had eight apiece. We have had of course and will continue to have losses - they dont shoot rubber bullets - but our boys will lick 'em. One B-17 fought off 15 Zeros and another 10 Zeros last week. In each case the B-17's were alone on reconaissance. Our gunners - all enlisted men - have plenty of guts and make the Japs pay for any attack they launch.

My first air raid was while I was in the New Guinea area. Bombs fell pretty heavily and pretty close. I crawled into a slit trench with the rest of the people - feeling foolish. One quickly gets over that feeling however. Just had dinner with a 2d Lt who is just out of the hospital. A zero got on his tail and shot him down. He bailed out at around 500 feet. Pulled out a dive of about 500 mph and managed to get out. These young pilots are plenty brave.

Letter from Dad

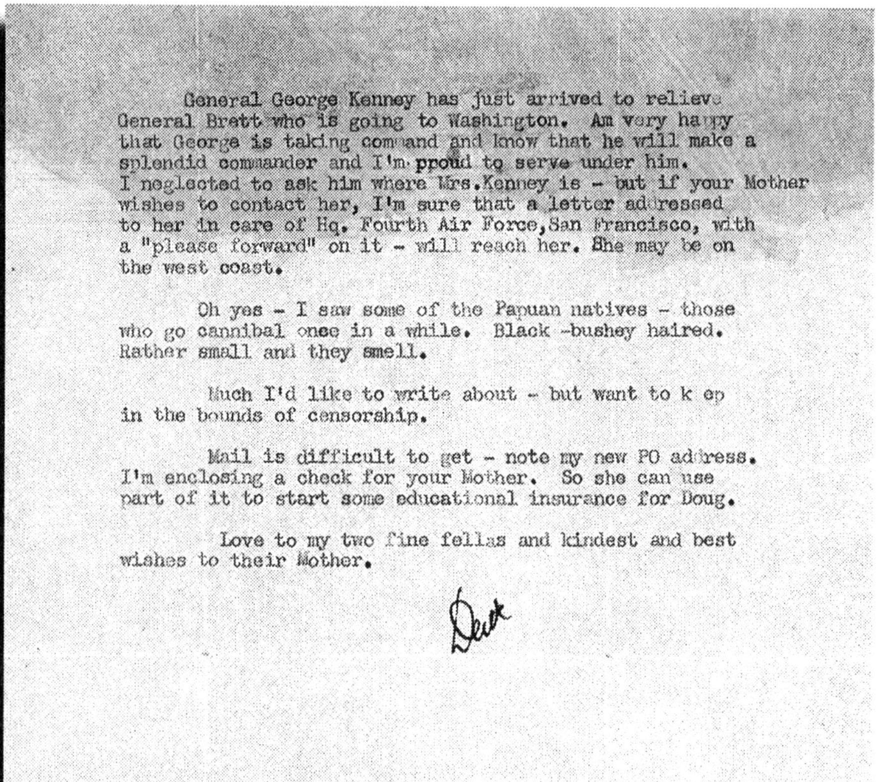

General George Kenney has just arrived to relieve General Brett who is going to Washington. Am very happy that George is taking command and know that he will make a splendid commander and I'm proud to serve under him. I neglected to ask him where Mrs.Kenney is - but if your Mother wishes to contact her, I'm sure that a letter addressed to her in care of Hq. Fourth Air Force,San Francisco, with a "please forward" on it - will reach her. She may be on the west coast.

Oh yes - I saw some of the Papuan natives - those who go cannibal once in a while. Black -bushey haired. Rather small and they smell.

Much I'd like to write about - but want to k ep in the bounds of censorship.

Mail is difficult to get - note my new PO address. I'm enclosing a check for your Mother. So she can use part of it to start some educational insurance for Doug.

Love to my two fine fellas and kindest and best wishes to their Mother.

Deuk

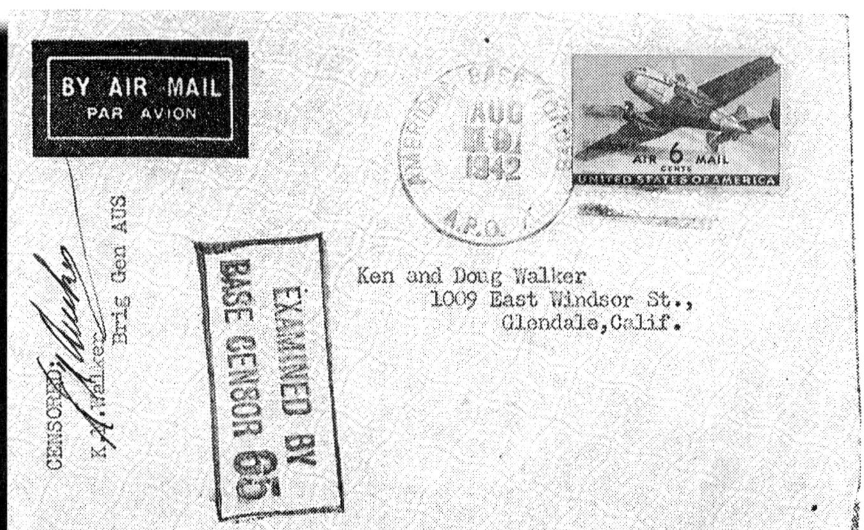

BY AIR MAIL
PAR AVION

AMERICAN BASE FORCES
AUG
1942
A.P.O.

AIR 6 CENTS MAIL
UNITED STATES OF AMERICA

CENSORED K.N. Walker
Brig Gen AUS

EXAMINED BY
BASE CENSOR 65

Ken and Doug Walker
1009 East Windsor St.,
Glendale,Calif.

COL. K. N. WALKER OF DENVER MADE BRIGADIER GENERAL

Brig. Gen. Willis H. Hale of Springs Becomes Temporary Major General.

Col. Kenneth N. Walker, 44, of Denver, was elevated to temporary rank of brigadier general in the army air corps by President Roosevelt Tuesday.

General Walker, reared and educated here, was commissioned a second lieutenant Nov. 2, 1918, and has remained in the air force ever since. On the occasion of his last visit here, Dec. 23, 1937, the heavy army bomber he was piloting was damaged while attempting a takeoff from the municipal airport. Walker, then a captain, and eight other army men in the plane narrowly escaped death.

Also advanced by presidential order was Willis H. Hale of Colorado Springs, from brigadier general to temporary major general. Brig. Gen. Ralph Royce of Hancock, Mich., who led the spectacular raid on Japanese positions in the Philippines this spring, received a similar promotion.

Maj. Gen. Joseph T. McNarney of Emporia, Pa., deputy chief of staff of the entire army, was advanced to temporary lieutenant general.

Delos C. Emmons, a temporary lieutenant general, was promoted to major general in the permanent rankings.

Other colonels who became brigadier generals were Edwin S. Perrin of Custer, S. D., Ennis C. Whitehead of Westphalia, Kan., Carl W. Connell of Leesburg, Va., Albert L. Sneed of Conway, Ark., and Nathan F. Twining of Oswego, Ore.

11th Bambardment Squadron

Brigadier General

NEMESIS OF BIG JAP SEA FORCE—Lieut. Gen. G. C. Kenney (center, first row), commander of Allied air forces in the Southwest Pacific, with assistants, fliers and plane crewmen of his organization. General Kenney directed and doubtless many of these men participated in the great air raid which wiped out the Jap convoy of ten warships and twelve transports trying to reach New Guinea. It was one of the great victories of this Pacific war. An American Flying Fortress forms the background for the photograph, made recently at an advanced bomber base, when many of the men shown were decorated for air exploits. To the left of General Kinney is Lieut. Col. R. H. Carmichael, and to the right of him, Brig. Gen. K. E. Walker—(Wirephoto).

Honored by MacArthur

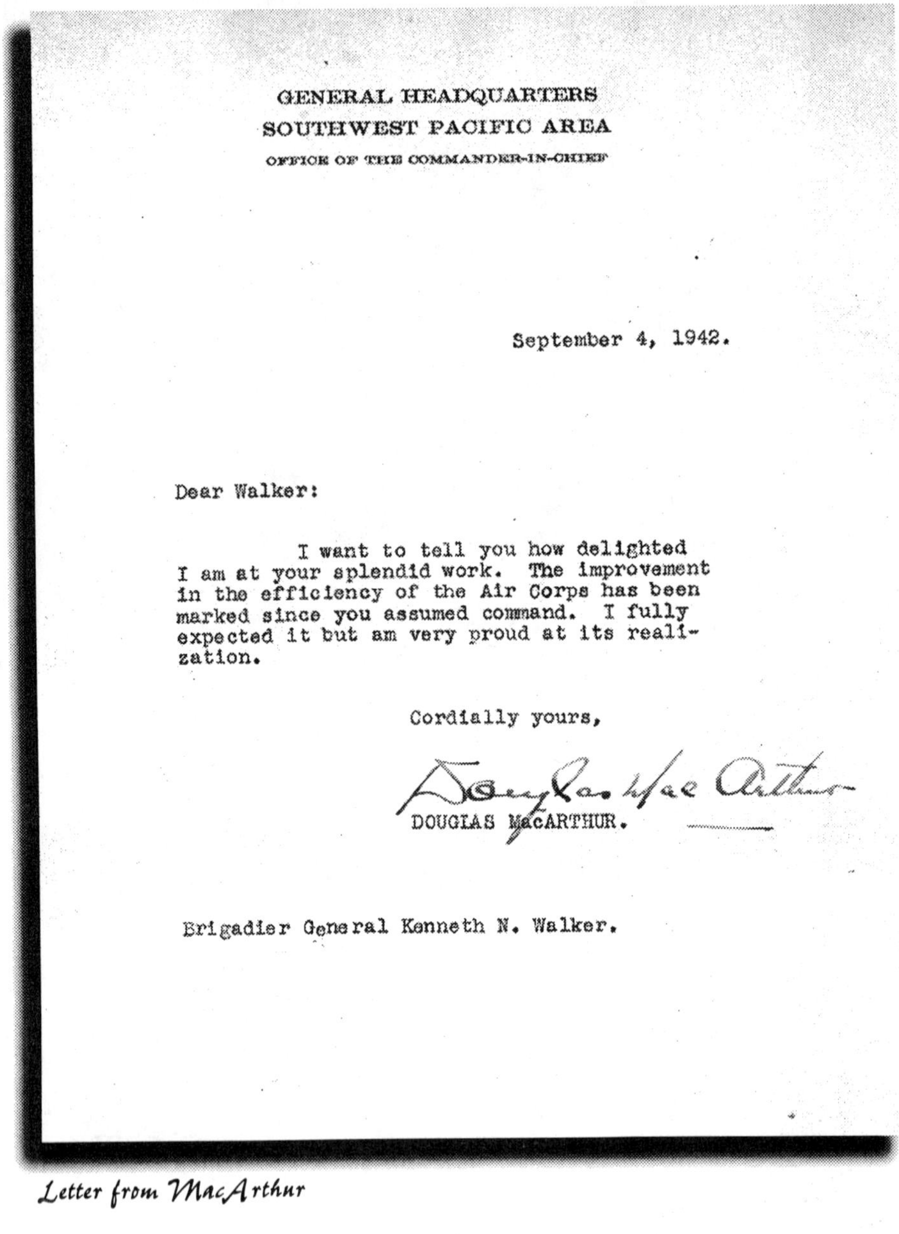

GENERAL HEADQUARTERS
SOUTHWEST PACIFIC AREA
OFFICE OF THE COMMANDER-IN-CHIEF

September 4, 1942.

Dear Walker:

I want to tell you how delighted I am at your splendid work. The improvement in the efficiency of the Air Corps has been marked since you assumed command. I fully expected it but am very proud at its realization.

Cordially yours,

Douglas MacArthur

DOUGLAS MacARTHUR.

Brigadier General Kenneth N. Walker.

Letter from MacArthur

ADDRESS REPLY TO
HEADQUARTERS OF THE ARMY AIR FORCES
WAR DEPARTMENT
WASHINGTON, D. C.

WAR DEPARTMENT
HEADQUARTERS OF THE ARMY AIR FORCES
WASHINGTON

October 19, 1942

Brigadier General Kenneth N. Walker
A.P.O. 500
C/O Postmaster
San Francisco, California

Dear Ken:

Today I saw a picture of us taken in the B-17 and it brought to my mind again that the days are slipping by and I have yet to thank you for the many thoughtful and considerate things you did to make the time I spent with you pleasant and purposeful in every respect.

Those who accompanied me as well as myself are most appreciative of your thoughtful arrangements that resulted in our having breakfast on our arrival and luncheon before our departure.

The day we were together was one of the most interesting and informative of the entire trip.

Because of your thorough description of your facilities and the inspection we made together, I feel I am much better acquainted with your needs and problems.

Again, many thanks for all you did to contribute to the success of our trip.

Sincerely yours,

H. H. Arnold

H. H. ARNOLD
Lieutenant General, U.S.A.
Commanding General, Army Air Forces

Letter from Arnold

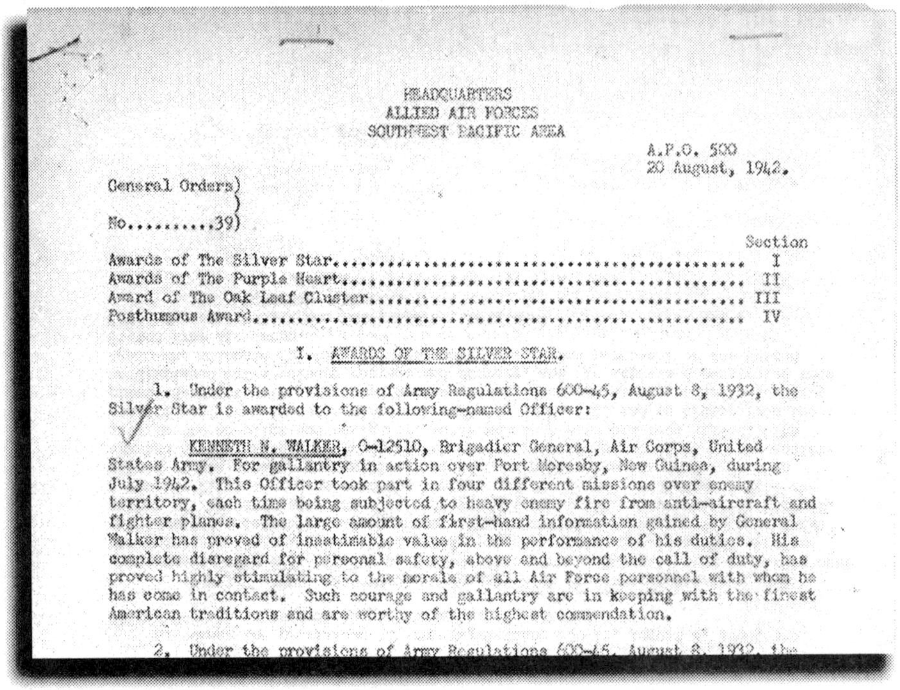

HEADQUARTERS
ALLIED AIR FORCES
SOUTHWEST PACIFIC AREA

A.P.O. 500
20 August, 1942.

General Orders)
)
No..........39)

	Section
Awards of The Silver Star..	I
Awards of The Purple Heart...	II
Award of The Oak Leaf Cluster..	III
Posthumous Award...	IV

I. AWARDS OF THE SILVER STAR.

1. Under the provisions of Army Regulations 600-45, August 8, 1932, the Silver Star is awarded to the following-named Officer:

KENNETH N. WALKER, O-12510, Brigadier General, Air Corps, United States Army. For gallantry in action over Port Moresby, New Guinea, during July 1942. This Officer took part in four different missions over enemy territory, each time being subjected to heavy enemy fire from anti-aircraft and fighter planes. The large amount of first-hand information gained by General Walker has proved of inestimable value in the performance of his duties. His complete disregard for personal safety, above and beyond the call of duty, has proved highly stimulating to the morale of all Air Force personnel with whom he has come in contact. Such courage and gallantry are in keeping with the finest American traditions and are worthy of the highest commendation.

2. Under the provisions of Army Regulations 600-45, August 8, 1932, the

Silver Star

Planning Session

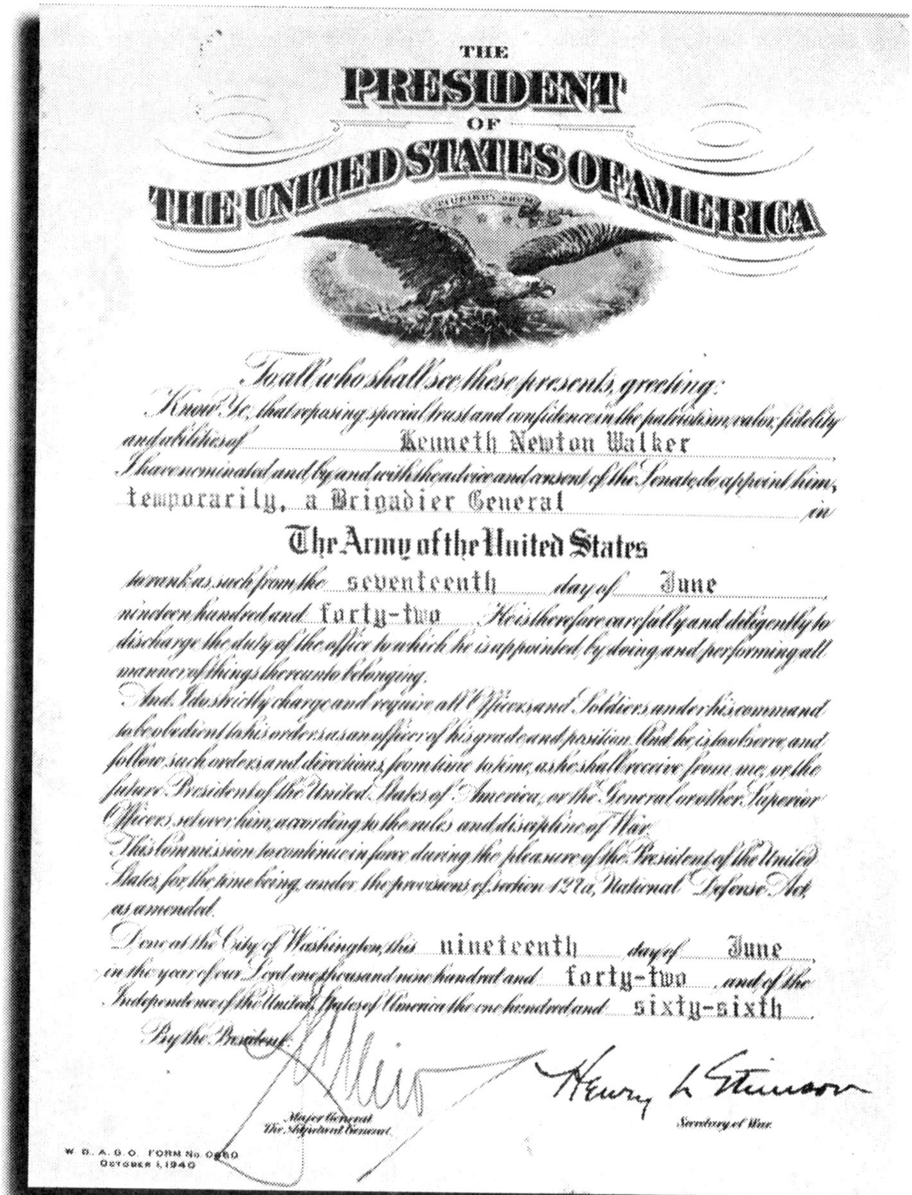

THE

PRESIDENT

OF

THE UNITED STATES OF AMERICA

To all who shall see these presents, greeting:

Know Ye, that reposing special trust and confidence in the patriotism, valor, fidelity and abilities of Kenneth Newton Walker I have nominated and by and with the advice and consent of the Senate, do appoint him, temporarily, a Brigadier General in

The Army of the United States

to rank as such from the seventeenth day of June nineteen hundred and forty-two. He is therefore carefully and diligently to discharge the duty of the office to which he is appointed by doing and performing all manner of things thereunto belonging.

And I do strictly charge and require all Officers and Soldiers under his command to be obedient to his orders as an officer of his grade and position. And he is to observe and follow such orders and directions, from time to time, as he shall receive from me, or the future President of the United States of America, or the General or other Superior Officers set over him, according to the rules and discipline of War.

This commission to continue in force during the pleasure of the President of the United States, for the time being, under the provisions of section 127a, National Defense Act as amended.

Done at the City of Washington, this nineteenth day of June in the year of our Lord one thousand nine hundred and forty-two, and of the Independence of the United States of America the one hundred and sixty-sixth.

By the President:

Major General
The Adjutant General

Henry L. Stimson
Secretary of War

W. D., A. G. O. FORM No. 0680
October 1, 1940

Presidential Citation

The Seattle Daily Times

MAin 0300 SEATTLE, WASHINGTON, MONDAY, SEPTEMBER 21, 1942.

General Roams Over Plane While His Boys Raid Japs

By Associated Press.

GENERAL MacARTHUR'S HEADQUARTERS, Australia, Sept. 21.—A young American general aims at flying with his boys against the Japanese at least once a week and shows he means business by going on 11 raids in less than two months. He is Brig. Gen. K. N. Walker, 44 years old, of Washington, D. C., whose wife and two sons, aged 14 and 9, live in the United States capital.

Boys All Like Him

"The boys in the north (of Australia) think the world of him," said the general's aide and pilot, Capt. Fred P. Dollenberg of Philadelphia.

"They figure things aren't so bad if a general's willing to go along and get shot at."

Carrying a bottle of oxygen, General Walker moves about a plane as it flies on its mission at a high altitude, Captain Dollenberg said.

"He climbs through the bomb bay and watches the rear gunner or the side gunners blast at Zeros and when we are over the target he watches the bombardier as he gets set to drop his bombs," he went on.

"Wandering all over a plane like that isn't healthy, but the general figures he can't tell the boys how to go out and to get shot at unless he's willing to get shot at, too."

General Walker, one of the youngest generals in the United States Army Air Forces, was in the War Plans Division in Washington before coming to Australia about three months ago.

The latest raid in which he took part was one against Rabaul, New Britain, deep in Japanese-occupied territory, last Friday night. Fires were started which were visible 30 miles away.

The general rode in a Flying Fortress on that trip. September 12 he was over Buna, New Guinea, in what was probably the heaviest raid of the Southwest Pacific area. On that occasion, Flying Fortresses, medium and attack bombers and fighters destroyed at least 17 Japanese planes and probably more on the ground.

One Mission a Week

"The general doesn't talk much about the raids," Dollenberg said. "But he figures he can't direct flights from the ground and tell the boys what they are doing wrong.

"So he goes along and directs a flight from the air. If a plane gets out of formation he shouts his orders over the radio to 'get the hell back in line.'

"The general figures on going on at least one mission a week. In less than two months he's been up with almost every squadron."

Willkie, in Moscow, plans frank talk with Stalin. Pg. 4.

Petrillo's ban hits industrial freedom—Arnold. Pg. 10

American Legion would draft manpower. Page 10.

The Daily Times

Flying General Goes on Raids

American Brigader, 44, Has Flown on 11 Forays in Less Than Two Months

GEN. MAC ARTHUR'S HEADQUARTERS (Australia) Sept. 21. (AP)—A young American general aims at flying with his boys against the Japanese at least once a week and shows he means business by going on 11 raids in less than two months.

He is Brig. Gen. K. N. Walker, 44, of Washington, D.C.

"The boys in the north (of Australia) think the world of him," said the general's aide and pilot, Capt. Fred P. Dollenberg of Philadelphia.

HE TAKES CHANCES

"They figure things aren't so bad if a general's willing to go along and get shot at."

Carrying a bottle of oxygen, Gen. Walker moves about a plane as it flies on its mission at a high altitude, Capt. Dollenberg said.

"He climbs through the bomb bay and watches the rear gunner or the side gunners blast at Zeros, and when we are over the target he watches the bombardier as he gets set to drop his bombs," he went on.

"Wandering all over a plane like that isn't healthy, but the general figures he can't tell the boys how to go out and to get shot at unless he's willing to get shot at, too."

ON RABAUL FORAY

Gen. Walker, one of the youngest generals in the United States Army Air Forces, was in the war plans division in Washington before coming to Australia about three months ago.

The latest raid in which he took part was one against Rabaul, New Britain, deep in Japanese-occupied territory, last Friday night. Fires were started which were visible 30 miles away.

He rode in a Flying Fortress on that trip. On Sept. 12 he was over Buna, New Guinea, in what was probably the heaviest raid of the Southwest Pacific area. On that occasion, Flying Fortresses, medium and attack bombers and fighters destroyed at least 17 Japanese planes and probably more on the ground.

Sutherland, Van Vaulkenburg, Walker

MacArthur, Walker, Kenney

Dollenburg, Overturf, Walker

General Walker in chow line

THE WASHINGTON POST: OCTOBER 18, 1942

D. C. Flier, 249 Others Get Valor Awards

Approximately 250 American air heroes, including Maj. Felix M. Hardison, of Washington, yesterday received a total of 386 medals for feats of valor performed during the United States Army's brief but colorful history in the southwest Pacific.

The presentations took place at a mass ceremony "somewhere in Australia" with Maj. Gen. George C. Kenney, commander of all Allied air forces in that area, doing the honors.

Major Hardison, who received the Silver Star on June 24 for heroism in a bombing flight which was intercepted by 12 Japanese Zero planes, yesterday was awarded three oak leaf clusters to go with his first medal.

The feat for which he was honored was not disclosed, but it was recalled here last night that war dispatches of August 29 named him as the leader of a bombing flight which battered a Japanese landing force at the southeastern tip of New Guinea, sinking a transport, probably sinking a cruiser and damaging a destroyer.

Among those who lined up yesterday to receive their awards—representing citations issued over a period of months—were privates, non-cimmissioned officers and commissioned officers. Seventy-five men received more than one decoration.

Outstanding in the latter group was 29-year-old Lieut. Col. Richard Carmichael of Olean, N. Y., who received the Distinguished Service Cross, the Distinguished Flying Cross and the Silver Star. The D. S. C. was for gallantry over Rabaul, New Britain, when Colonel Carmichael's flight knocked 7 of 11 Japanese planes out of the sky and destroyed or damaged 50 planes aground.

General Kenney, traveling in a Flying Fortress, was accompanied to the scene of the presentations by Brig. Gen. Kenneth E. Walker, of 1701 Massachusetts Avenue, Northwest, Washington. Walker figured in a story from Australia on September 21 "as a young American general who had gone along with his boys on 11 air raids in less than a month."

All of yesterday's awards could not be presented—some of the recipients were dead, some were hospitalized, others were away on war duty. And in no case was it announced what the awards were for, although previous dispatches made it clear why many of the group were decorated.

THE EVENING STAR, WASHINGTON, D. C., SATURDAY, OCTOBER 17, 1942.

386 Medals Awarded American Air Heroes At Australian Base

Maj. Felix M. Hardison Of Washington Among Those Honored

By MURLIN SPENCER,
Associated Press War Correspondent.

SOMEWHERE IN AUSTRALIA, Oct. 17.—Three hundred and eighty-six medals were pinned on American air heroes at an advance operational base today in the greatest mass presentation of awards for valor in the United States Army's brief but colorful history in the Southwest Pacific.

For two hours we watched privates, non-commissioned officers and commissioned officers move forward under a broiling tropical sun to receive their decorations from Maj. Gen. George C. Kenney of Gardiner, Me., commander of all Allied air forces in this area.

Approximately 250 men received the awards, representing citations issued over a period of months, with 75 receiving more than one decoration. The awards included Distinguished Service Crosses, Silver Stars, Distinguished Flying Crosses, Purple Hearts and Soldier's Medals.

The ceremony was staged on a wide airdrome runway hacked out of the jungle, with rows of huge Flying Fortresses forming a grim background.

Many Multiple Awards.

Among the outstanding winners of more than one decoration was Lt. Col. Richard Carmichael, a handsome, bronzed flyer of 29 from Austin, Tex., who received the Distinguished Service Cross, the Distinguished Flying Cross and the Silver Star.

Lt. Carmichael won his DSC for gallantry over Rabaul, New Britain, when his flight blasted 7 of 11 Japanese fighters out of the skies and destroyed or damaged 50 planes aground. His wife lives in Olean, N. Y.

The DSC also went to Master Sergt. Alva S. Huscall of Redlands, Calif.; First Sergt. Robert R. Davis of Albuquerque, N. Mex., and Master Sergt. William H. Prince of Santa Rosa, Calif.

The most decorated enlisted men were Sergt. Russell I. Huffman, Highwood, Mont., and Technical Sergt. Clevis O. Jones of Chatham, Va. Sergt. Huffman received the Silver Star and three oak leaf clusters in lieu of additional silver stars, and also the Purple Heart. Sergt. Jones received the DFC, Silver Star with oak leaf cluster and the Purple Heart.

Washingtonian Decorated.

Other multiple winners included: Second Lt. Eddie W. Hayman, Huntsville, Tex., DFC, Silver Star and Purple Heart; Technical Sergt. Arthur L. Richardson, Denver, Colo., DFC, Silver Star and Purple Heart; Maj. Felix M. Hardison, Washington, D. C., three oak leaf clusters to Silver Star; Capt. Kenneth D. Casper, Humboldt, Kansas, Silver Star, oak leaf cluster, DFC; First Lt. Curtis J. Holdridge, Atlas, Okla., Silver Star, oak leaf cluster, DFC; Staff Sergt. Stanley Jackola, Portland, Oreg., Silver Star, oak leaf cluster, DFC.

The awards represented action over a wide battle area, including Australia, New Guinea, New Britain, the Solomon Islands and the Philippines.

A Flying Fortress piloted by Capt. Wilbur Beesley of Syracuse, Nebr., carried Gen. Kenney, Brig. Gen. Kenneth E. Walker of Washington, D. C., and a big cardboard cigarette box loaded with the medals to the field for the ceremony.

Some of the medals could not be presented, for some of the heroes were casualties—dead or hospitalized by injuries. Others were on war details away from the field.

There was no statement as to what the recipients had done to win the awards, but that was not needed because their comrades who witnessed the ceremony and the group of nurses and other women seated in the shade of a bomber's wing knew. United States newspapers already had carried stories from Gen. Douglas MacArthur's headquarters announcing most of the details.

Award for Valor

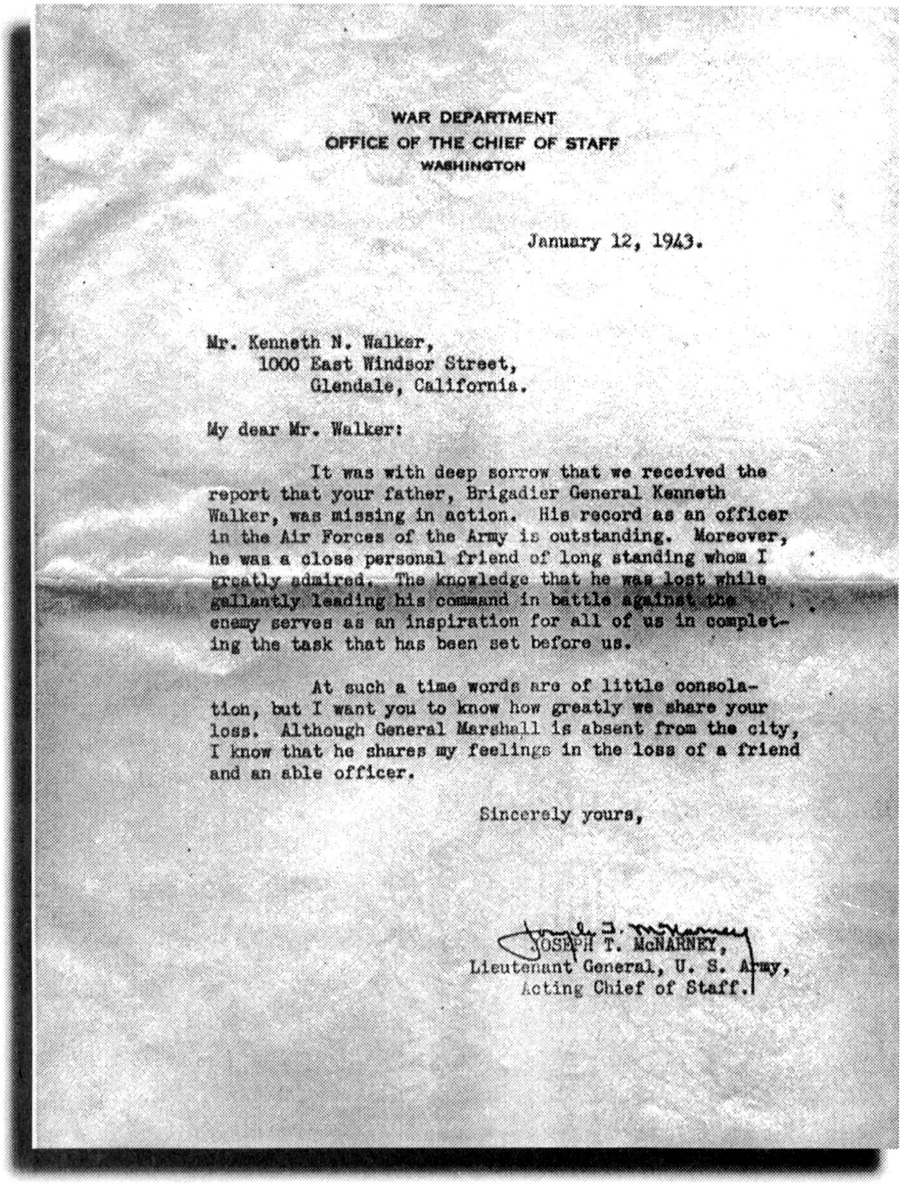

WAR DEPARTMENT
OFFICE OF THE CHIEF OF STAFF
WASHINGTON

January 12, 1943.

Mr. Kenneth N. Walker,
1000 East Windsor Street,
Glendale, California.

My dear Mr. Walker:

It was with deep sorrow that we received the report that your father, Brigadier General Kenneth Walker, was missing in action. His record as an officer in the Air Forces of the Army is outstanding. Moreover, he was a close personal friend of long standing whom I greatly admired. The knowledge that he was lost while gallantly leading his command in battle against the enemy serves as an inspiration for all of us in completing the task that has been set before us.

At such a time words are of little consolation, but I want you to know how greatly we share your loss. Although General Marshall is absent from the city, I know that he shares my feelings in the loss of a friend and an able officer.

Sincerely yours,

Joseph T. McNarney
JOSEPH T. McNARNEY,
Lieutenant General, U. S. Army,
Acting Chief of Staff.

Missing in Action

GLENDALE NEWS PRESS

Glendale Brigadier General Missing After Raid on Japs

Failing to return from a successful air raid on Rabaul, New Britain, on Jan. 5, Brig. Gen. Kenneth N. Walker of Glendale has been listed by Gen. Douglas MacArthur as missing in combat. "This officer," Gen. MacArthur said, "with the greatest personal courage, led a bombardment group which successfully attacked enemy shipping in Rabaul harbor. In this attack from nine to 11 enemy ships were destroyed."

It was reported from Allied headquarters in Australia that Flying Fortresses and Liberators damaged more than 50,000 tons of

BRIG. GEN. KENNETH N. WALKER, of Glendale, reported missing since Allied raid on New Britain.

shipping in the harbor ,and shot down nine Japanese Zeros.

"Gen. Walker's plane was last seen operating over the harbor and has not returned," Gen. MacArthur's statement said. "It must be presumed that he is lost.

"This officer's record in this campaign has been outstanding and much of the efficiency of the bombardment command of the 5th American air force is due to his exceptional brilliance and courage."

Gen. Walker was among a number of American and Australian officers recently cited by Gen. MacArthur to receive the distinguished service cross for "extraordinary courage, marked efficiency and precise execution of operations during the Papuan campaign."

Mrs. Walker, former wife of Gen. Walker, who resides at 1009 East Windsor with her two sons, Kenneth Walker jr., 15, and Douglas Walker, 9, expressed confidence this morning that the general will be found safe.

Third General Lost

"Kenneth has been in difficult situations before and came out safely," she said. "Several times he has had to make forced landings in bombers and has always come out without being hurt."

Gen. Walker is the third American air force general lost on combat missions in this war. Maj. Gen. Clarence F. Tinker has been missing since leading a bomber force against the Japanese in the battle of Midway and Brig. Gen. Asa N. Duncan has been missing since November when he left England for a flight to North Africa.

Walker Missing

—News-Press Photo

KENNETH N. WALKER JR., of Glendale, who is en route to Washington to receive from President Roosevelt a medal of honor awarded to young Walker's father, Brig. Gen. Kenneth N. Walker, reported missing in action.

BRIG. GEN. KENNETH N. WALKER, who has been awarded a medal of honor for conspicuous gallantry at Rabaul, New Britain.

General's Son To Be Honored

Honor and distinction that come to few persons is the lot of Kenneth N. Walker jr., 16-year-old Glendale youth, who is en route to Washington to receive from the hands of President Roosevelt a medal of honor awarded to young Walker's father, Brig. Gen. Kenneth N. Walker for "conspicuous gallantry at Rabaul, New Britain. Gen. Walker was reported as missing in combat after his failure to return from a successful air raid on Rabaul last Jan. 5.

According to a telegram received from the adjutant general's office in Washington, young

—Turn to Page 2, Col. 4—

—Continued from Page 1—

Walker has been invited to come to Washington and receive the award as Gen. Walker's next of kin. President Roosevelt has set aside 11:15 a. m., Thursday, March 25, as the time for the ceremony at the White House.

Notification of the honor awarded his father was contained in the following letter from the war department:

"Kenneth N. Walker Jr.,
"1009 East Windsor Road,
"Glendale, Calif.

"Dear Kenneth:

"It gives me great pleasure to inform you that your father, Brig. Gen. Kenneth N. Walker, has been awarded the Medal of Honor, the senior military decoration of the United States. This medal supersedes the previous award of the Distinguished Service Cross and is a fitting tribute to a gallant officer.

Limited Number Given

"In order that you may better understand the honor of this decoration, I would like you to know that there has been a very limited number awarded in the entire history of our country. They are given only for 'gallantry and intrepidity at the risk of life above and beyond the call of duty.' Your father was indeed deserving of this recognition.

"Gen. Walker has brought honor to the entire army air forces, and we are proud that his name will be recorded with the heroes of American history.

"Very sincerely,
"GEORGE E. STRATEMEYER,
Major General, United States Army, Chief of the Air Staff."

'Extraordinary Courage'

Gen. Walker was among a number of American officers cited by Gen. MacArthur to receive the Distinguished Service Cross for "extraordinary courage, marked efficiency and precise execution of operations during the Papuan campaign." His plane was last seen operating over Rabaul harbor on Jan. 5 and has not returned," Gen. MacArthur's report said at that time. "It is presumed that he is lost."

Since that time, however, the general's wife, who lives with her two sons, Kenneth and Douglas, 9, at the Windsor road address, has received unofficial information that Gen. Walker, with several of his command are being held as Japanese prisoners.

Aided by Legion

Arrangements for young Walker's trip to Washington were made by officers of Glendale post, No. 127, the American Legion, which also provided funds for incidental expenses of the trip.

Cmdr. Earl E. Baldwin advised Sen. Hiram Johnson (R-Calif) and Rep. Carl Hinshaw (R-Calif) that the Glendale boy will be at the White House for the presentation ceremony and asked Rep. Hinshaw to represent the Glendale Legion post officially. Cmdr. Baldwin also sent a letter of introduction to Washington and another letter to President Roosevelt expressing appreciation of the Glendale Legion post.

Ken Walker Jr. honored

Marguerite Potter Walker with sons Ken Jr. and Doug

Appendix B

Driving Home the Bombardment Attack

By 1st Lieut. K. N. Walker, A. C.

General

When a bombardment unit clears its airdromes with a mission of destroying a vital objective deep within a hostile territory, it will be opposed vigorously by the enemy's defense forces; the hostile pursuit aviation and antiaircraft artillery. The unit will be confronted with a task no more difficult than that which confronts the infantry when it jumps off on a well-planned and coordinated attack. As the infantry receives the support of other ground arms, so does the bombardment unit receive the support, either special or general, from the other classes of aviation—observation, attack and pursuit, necessary to drive home the bombardment attack. In examining the tactics which a bombardment unit will employ to insure its arrival over and the attack of the objective, it will be found that this class of aviation operates at high altitudes and low; by day and by night; in formation and by series of single airplanes.

A brief review of bombardment organization and equipment may assist somewhat in the understanding of the present accepted bombardment tactics. Bombardment aviation is organized into squadrons, groups and wings. Two or more combat squadrons with a service squadron and other auxiliary units compose a group; two or more groups with necessary auxiliary units compose a wing. Each squadron is equipped with thirteen airplanes, of which a maximum of ten are expected to be always in commission. The group, with four combat squadrons, for example, is expected to put a maximum of forty airplanes in the air.

The number of squadrons which will operate against a particular target will depend upon the type of objective and the hostile opposition expected. It is impossible to determine,

---

Walker, Kenneth N., "Driving Home the Bombardment Attack," *Coast Artillery Journal* 73, no.4 (October 1930): 328–40.

in the abstract, the strength which must be employed against any particular type of target. It will, to use the overworked phrase, "depend upon the situation." However, bombardment tactics are developed with a view to the proper employment of whatever number of airplanes must be used to accomplish a mission, rather than being based upon specific types of objectives which bombardment aviation will attack. The tactics developed are adapted primarily to the squadron and group organization. When more than one group is employed against a particular objective, the wing tactics consist of one group guiding upon the other, maintaining such intervals as are necessary for coordinated action and mutual support.

Bombardment aviation employs two types of airplanes; the heavy bombardment airplane capable of carrying a two thousand four hundred-pound bomb load; the light bombardment airplane capable of a one thousand two hundred-pound bomb load. Ordinarily the units equipped with the heavy bombers will operate at night. Those equipped with the light bombers will operate in daytime. Each, however, are suitable for and may operate both day and night. While at the present time the heavy bombardment airplanes only are in service use, light bombardment airplanes are under construction. The heavy bomber must be capable of high speed of at least one hundred and twenty-five miles per hour and must have a radius of at least three hundred miles; the light bomber a high speed of one hundred and sixty miles per hour and a radius of action of two hundred miles. Each bomber is twin-engined—one engine placed out-board on either side of the fuselage. In the nose of the fuselage extending forward of the leading edge of the wings, are placed the pilot, the bomber and the front gunner. In the rear portion of the fuselage are located the rear gunners. An alternate arrangement of rear gunners is to place them in the engine nacelles. By such an arrangement, excellent vision for the pilot and the bomber are afforded. Flexibly mounted machine guns cover all areas open to the approach of hostile attacking aircraft. The airplane is equipped with radio telephone with which communication between airplanes may be maintained and formations controlled in flight. With the rapid advance of aeronautical development the above conception of the proper types of bombardment airplanes will be changed from time to time, when increased performance will make possible greater loads and cruising range.

For purposes of discussion, bombardment tactics will be reviewed under the following headings:

Day operations at high altitudes.
Day operations at low altitudes.
Night operations.
Special support by other classes of aviation.

Day Operations at High Altitudes

In conducting day operations at high altitudes—meaning altitudes above ten thousand feet—a bombardment unit will normally perform a mission with its airplanes in formation. The formation lends itself to the delivery of a mass attack, to defensive machine gun fire superior to that which may be brought against it and affords a measure of security against antiaircraft fire. That a mass attack is delivered from a formation is, of course, obvious. It is necessary, however, to investigate the types of formations adopted to understand their defensive powers when opposed to hostile pursuit and antiaircraft artillery.

Although it is not desired to consider in detail all points concerning the bombardment formation, it is believed that the discussion which follows will indicate sufficiently the features upon which the foregoing statements are predicated. First, the formation must be simple, compact and capable of ready control by the formation leader. It must be capable of maneuver and so flexible that distances and intervals between individual airplanes may be readily opened and closed. Its arrangement must be such that all angles of approach by hostile aviation are well covered by defensive machine gun fire. In this connection, emphasis is placed upon a formation arrangement whereby the maximum fire may be concentrated against that angle most favorable to attack by single seater pursuit. The formation must be so flown that a simultaneous attack by a superior number of hostile pursuit is difficult.

To meet these requirements, the normal formation consists of a number of three or five airplane elements. Within elements the airplanes are echeloned rearward from the leading airplane to the right and left and slightly upward in altitude, forming a V. Each element flies to the rear of the preceding element. The elements are echeloned *downward* from front to rear. With such arrangement all areas enclosing the formation are well covered by machine gun fire. By the "staggered down" feature embodied in echeloning elements downward from front to rear, all rear gunners are provided with unblanketed fire to the upper rear hemisphere, which is the angle of approach most favorable to the attack of hostile single seater pursuit. To appreciate the fact that approach from the upper rear hemisphere is most favorable to hostile pursuit consider the difficulties of the frontal or flank attack, or the attack from the lower rear hemisphere of a formation. In the frontal attack, the speed of approach of the pursuit is the sum of the speeds of the pursuit airplanes and the bombardment formation. This great speed limits the time in which the attacking pursuit is in position to deliver accurate aimed fire to but a few seconds. In the flank attack the target is moving at right angles to the line of fire of the pursuit—aimed fire is again difficult. In the attack from the lower rear hemisphere, the pursuit airplane pulls up from a dive beneath the formation. The speed of the pursuit airplanes is materially reduced in the upward climb and the airplane "hangs" beneath the formation within range of the bombardment machine guns a longer period of time than is available for the pursuit airplanes to deliver aimed fire. In an attack from the upper rear hemisphere, the speed of approach is the difference between the speed of the formation and the attacking airplane; the front guns of the latter may be aimed from the beginning of the dive to completing the attack; the speed built up in the dive insures rapid withdrawal upon completion of the attack; the formation is moving generally in line of fire of the attacking pursuit airplane. Pursuit will attack from all angles, however, and as above noted, all angles of approach are well covered by machine gun fire, but with the maximum gun fire available to the upper rear hemisphere.

As the upper rear hemisphere is most favorable to pursuit attack, the formation is as narrow laterally as is consistent with concentration of defensive machine gun fire, to make difficult the simultaneous attack by large numbers of pursuit airplanes from this angle. Thus, a bombardment group formation of four squadrons can be easily flown within an area five hundred feet wide and one thousand feet long. It will be most difficult for equal or superior number of pursuit airplanes to launch a coordinated, concentrated attack against a group formation of this character. Even though forty pursuit airplanes could deliver a simultaneous attack against such a formation, it would be bringing but eighty machine guns into action, against either one hundred and sixty or two hundred and forty guns mounted on forty bombardment airplanes. As the rate of fire of the flexibly mounted machine gun is nearly twice that of the machine gun mounted to fire through the propeller, it is apparent that the bombardment formation should have the best of the argument, by sheer force of fire power.

Pursuit will attack by long range fire, as well as by close range fire. A group formation as compact as that above discussed is undesirable, in that machine gun fire, delivered in the plane of the formation, may miss the airplane at which aimed, but hit another airplane. When such fire is anticipated the bombardment formation may increase interval and distance between airplanes to from one to two hundred feet in from about one to three seconds. A hostile pursuit force may attack with a number of airplanes flying to the rear of the formation and delivering long range fire, while other airplanes deliver close range fire, approaching the formation from several angles. The open formation will be used against such an attack. Accuracy in fire will be an important factor in the relative number of pursuit and bombardment airplanes hit. The fact remains that the bombardment formation is still delivering a superior volume of fire against the attackers.

Pursuit may employ a time-fuzed fragmentation bomb which may be dropped on a bombardment formation from above. Two-seater pursuit is being developed. A hostile force equipped with airplanes of this type, may form on the flanks and in front of a bombardment formation, and concentrate against it the fire of the flexibly mounted rear guns. While the bombardment formation is the recipient of either or both of the above types of attack, other pursuit may attack the formation from the rear with their fixed guns. For the defense against this type of attack, support by friendly pursuit may be required.

A formation designed for defense against hostile aircraft is not entirely suitable for the avoidance of antiaircraft gun fire. A compact defensive formation is less maneuverable and it provides a larger target against which all antiaircraft batteries within range may be concentrated. That formation most suitable for operations over areas defended by antiaircraft artillery, consists of one in which the airplanes are flown with considerable intervals and distances, i. e., where the airplanes are dispersed rather than concentrated. One type of dispersed group formation, known as the "dispersed column" is cited to illustrate. In this formation each squadron will have ten airplanes, the normal number. The airplanes are flown in two elements of five airplanes each, one behind the other. When opening up to a dispersed column, the leading squadron maintains the lower altitude. The second and third squadrons take positions on the flanks and above to the rear of the leading squadron, each maintaining a distance of approximately one thousand five hundred feet from the leading squadron. The fourth squadron flies to the rear of the leading squadron at a distance of approximately three thousand feet and about two thousand feet above the leading squadron. Within squadrons, the second elements are echeloned upward in altitude to the rear of the first or leading element. The individual airplanes are flown from four hundred to six hundred feet apart in their respective elements. Within such a formation, the airplanes are constantly changing speed, altitude and direction in maintaining the assigned distances. When antiaircraft fire is anticipated or experienced, each airplane, guiding upon the one in front of it within its respective element, engages in decided maneuvers. Endeavor is made to change altitude, speed or direction, or a combination of these, within the time of flight of the antiaircraft shell to the altitude at which the airplanes are flying. With these distances between airplanes, one antiaircraft shell can injure but one airplane. Should all batteries within range concentrate on one squadron, the other squadrons are not in danger. If the batteries do not concentrate their fire, the probability of hits is reduced. When attacking a compact defensive formation, all batteries may concentrate their fire against *the formation*, with the probability that slight errors in fire, directed against a particular airplane will hit another airplane in the formation, and that a shell which hits or detonates near one airplane may seriously damage another airplane.

In a group formation such as described above, forty individual and separate targets are presented to the antiaircraft artillery. By plotting an antiaircraft gun defense, the area in which effective fire may be delivered is of course determined. The time during which the formation will be within range of the batteries may be calculated. A formation flying at a speed of from two to three miles a minute will be within effective antiaircraft range but for a short space of time. These tactics present a problem to the defending antiaircraft artillery far greater than that presented when the bombardment formation approaches an objective in a compact formation.

In bombing from such a formation, each airplane is held to a straight course for those seconds (not to exceed twenty) required to perform the timing operation and release the bombs. Upon release of bombs the airplanes again assume a maneuvering course until the defended area is passed through.

When a bombardment unit takes off to perform a mission, it will normally open to a dispersed formation. The compact defensive formation is required only for defense against hostile aircraft. The bombardment pilots are subjected to less strain in flying the dispersed formation than in the defensive formation. The route selected for the mission will avoid, as nearly as possible, the known or suspected areas in which hostile pursuit is certain to be operating, and where antiaircraft artillery is sure to be emplaced. A route around open flanks will be preferred to a route which requires the formation to cross the combat zone of the enemy, wherein hostile pursuit is certain to be operating, and where antiaircraft artillery is sure to be emplaced. If, however, there is no alternative, the formation will proceed to the objective through the hostile combat zone. As it is possible that the hostile pursuit will attempt to intercept the bombardment formation as it crosses the hostile front lines, the air force commander will arrange that, at the time and place where the bombardment unit crosses the lines, friendly pursuit will be present in force. By such action, the bombardment formation may be enabled to maintain the dispersed formation while flying over the combat zone. Should hostile pursuit be present, the friendly pursuit should be able to effectively prevent the former from attacking the bombardment formation. A combat zone of twenty to thirty miles in depth may be crossed in from ten to fifteen minutes. Unless the enemy pursuit succeeds in engaging the bombardment formation without being prevented by the friendly pursuit the bombardment should be able to traverse the combat zone in dispersed formation and thus limit the effect of antiaircraft opposition. If, however, the hostile pursuit is present in force and is not prevented from attacking the bombardment formation, the latter will assume a defensive formation. It is unlikely that hostile antiaircraft will fire when its own pursuit is present and engaged in attacking the bombardment formation. A coordinated attack by pursuit and antiaircraft would be difficult of accomplishment without considerable danger to the pursuit. The antiaircraft doctrine, which in effect is that when friendly pursuit is present in force, the antiaircraft artillery withholds its fire, is logical and will doubtless be applied.

Upon passing through the combat zone, the bombardment route will avoid the antiaircraft batteries grouped around vital points in the system or rail communications, important supply establishments, etc. Should hostile pursuit be absent or prevented by friendly pursuit from attacking the bombardment formation when the bombardment formation crosses the hostile front, it is expected that hostile pursuit units in the air and on the alert at airdromes will be notified of the presence of the bombardment formation. These pursuit units will endeavor to intercept the bombardment formation as quickly as possible. The time required for interception will be that necessary for transmission of information to the pursuit units; the time required to issue orders, clear the airdrome and climb to the altitude at which the bombardment formation is flying, if the pursuit unit be on the ground; the difference in speeds between the bombardment formation and friendly pursuit; and the accuracy with which the hostile pursuit units carry out the interception. Should interception by hostile pursuit be effected before the objective is reached, the bombardment unit will assume a defensive formation and engage in a running fight until the objective is reached. If the hostile pursuit force fails in preventing the bombardment formation from reaching its objective, but continues attacking the latter when it arrives within range of the antiaircraft artillery, the bombardment unit will assume a dispersed formation, release its bombs therefrom, and be prepared to close up to a defensive formation upon clearing the range of the antiaircraft batteries. The time required for a bombardment formation to open or close, to assume one formation or another, is measured in seconds, rather than minutes. It may appear wise in theory for the antiaircraft to refrain from firing when a dispersed formation is assumed, on the proposition that pursuit will then dive in to engage the bombardment airplanes, or for pursuit to refrain from attacking a defensive formation on the assumption that antiaircraft artillery will then open fire. However, it is submitted that the extreme nicety of coordination of such tactics will cause delays which, measured in time, will be such to allow the bombardment formation to proceed a great deal of the time without being subjected either to the fire of pursuit or antiaircraft. If the action of pursuit and antiaircraft can be so coordinated and perfected that antiaircraft can fire during the intervals between successive pursuit attacks, without danger to the attacking pursuit, it may be then habitual for friendly pursuit to support a bombardment formation. Friendly pursuit should be able to break up any coordinated attack by the hostile pursuit, thus permitting the bombardment unit to maintain an open or dispersed formation without sacrifice of the scheme of defensive machine gun fire.

Day Operations at Low Altitudes

Considerable thought, in both this and in foreign countries, is being devoted to the employment of low altitude tactics for bombardment aviation. The tactics will closely parallel those now adopted by attack aviation. Altitudes flown will vary between the tree tops and five hundred feet. When flying over areas definitely known to be unoccupied by troops of any kind, this altitude may be increased to aid navigation.

The employment of low altitude tactics is not intended to limit, in any way, the function of bombardment aviation. Its normal objectives will continue to be those material objectives outside the radius of action or beyond the power of destruction of attack aviation, and outside the range of artillery. Low altitude tactics is but another method by which bombardment units will carry out their missions.

Missions will be conducted at low altitudes when the weather conditions are such that a limited ceiling obtains. Low altitude operations will also be alternated with high altitude operations to bring about that uncertainty which will prevail when the enemy is doubtful as to the manner in which the bombardment attacks will be conducted.

While a formation similar to the defensive formation outlined for high altitude operations can be flown at extremely low altitudes, it is believed that for low altitude operations, the bombardment formation most suitable is one similar to that flown by attack aviation. In this formation, each squadron employs nine airplanes divided into three, three airplane elements. The second and third elements are echeloned to the rear and to the right and left of the leading element and at the same altitude. Thus, each squadron will form a V of three airplane elements. When two or more squadrons are employed on a mission, the squadrons follow the leading squadron in column, at distances of approximately one thousand feet between squadrons.

When conducting a mission, the route selected will avoid particularly areas occupied by hostile ground troops, to evade rifle, machine gun and .37-mm. gun fire. Upon approaching the objective, if it be a precision target, the elements will close upon the leading element in each squadron. The squadrons will either close up to distances of two hundred feet between squadrons, or will increase their distances to approximately two thousand feet. The first method may be employed when the bombs are equipped with delay fuzes of from ten to fifteen seconds. The latter method may be employed when the bombs are equipped with five second delay fuzes. By the first method detonation will occur when the entire formation is beyond the danger radius of the bombs. By the second method, the bombs dropped by one squadron will detonate after it has passed beyond the danger radius of the bombs and before the following squadron is within that danger radius.

If the formation is intercepted by hostile pursuit, the elements within squadrons close upon the leader, maintaining their original echelonment, or taking position in column behind the leader. The squadrons close to at least five hundred feet between each other.

In peacetime maneuvers, considerable difficulty has been experienced by hostile pursuit in intercepting low flying formation, even with the wings of military airplanes painted a bright yellow. When the wings and fuselage are well camouflaged, it is believed that it will be practically impossible for pursuit pilots at high altitudes to see a formation flying just above the three tops. A great deal of assistance will be required from ground agencies to aid hostile pursuit in effecting interception.

It may become necessary for pursuit to maintain units both at extremely high altitudes and at altitudes of around five thousand feet, to defend against bombardment units which may employ either type of tactics. Obviously an additional burden is then placed upon the defending pursuit. The engines of pursuit airplanes, supercharged for good performance at high altitudes, do not give equal performance at low altitudes. It may be necessary to equip part of a defending pursuit force with engines designed to give the best perception, the bombardment unit will close up to a defensive formation. The hostile single-seater pursuit will be restricted in its maneuvers. If two-seater pursuit intercepts and surrounds a bombardment formation, and if the methods of bombing the formation and attacking from the rear, as described previously, are used, the situation will be similar to that which confronts the formation flying at high altitudes. If interception of this type of hostile aircraft proves to be the rule rather than the exception, supporting friendly pursuit may be necessary.

Although increased accuracy than that now secured is expected from antiaircraft guns firing at low angles of elevation, it is believed that the problem of effectively ranging upon a low altitude formation will be always greater than against a formation at high altitudes. The greatest menace to the unit employing low altitude tactics will be that of hostile rifle, machine gun and .37-mm. gun fire. A great deal of this fire may be avoided in selection of the route which the formation will follow to the objective. At the bombardment objective, it is expected that in addition to antiaircraft gun defense, a system of machine gun defense, and even rifle defense may be established. Obviously, if bombardment aviation employs both high and low altitude tactics, an additional burden is placed upon the antiaircraft which defends a bombardment objective, requiring both guns and machine guns. Greater surprise is ordinarily obtained in the low altitude attack than in the high altitude attack. The defending forces must be ever on the alert to meet an attack from either high or low altitudes.

Night Operations

Night attacks against a bombardment objective will be conducted by individual airplanes, rather than by large formations. This does not mean that but one airplane will attack an objective at one time, followed by another at a considerable interval. On the contrary, three or more individual airplanes will attack simultaneously, from different directions and at different altitudes. Predicated upon the assumption that the single airplane will be illuminated prior to its arrival over that point where its bombs will be released, it is believed proper to employ tactics which will lessen the chances of such illumination. Whether or not the illuminated airplane is hit by a shell fragment, the bright light in the bomber's cockpit will affect adversely precision bombing. Colored filters for the cockpit and colored goggles for the bomber will eliminate the glare somewhat, but the illuminated airplane will have less chance of performing accurate bombing than one not illuminated.

A night operation which might be performed by a group is outlined to illustrate briefly the method used in a night attack. In conducting the group mission, each squadron is ordered to make three attacks at five-minute intervals. Each attack is to be made by three airplanes. The time of the delivery of the first attack of each squadron will be prescribed. Three routes will be prescribed by the group commander between the airdromes and the objective. The latter commander will also specify the directions of approach to the objective and the altitudes from which the airplanes, following each route, will approach. Upon release of bombs each airplane will make a one hundred and eighty-degree turn, diving to a lower prescribed altitude and leave the objective by the same route which was followed to the objective. With the exception of those airplanes making the first and last attacks on each of the three routes, one airplane will be approaching the objective at one altitude, while another airplane will be leaving the objective at a lower altitude. The two airplanes should cross at a point where the antiaircraft is endeavoring to locate and illuminate the incoming plane.

By such tactics a more difficult problem is presented to the defending antiaircraft artillery, than if the bombardment airplanes approached the target, one at a time. If all gun and searchlight batteries concentrate on one bombardment airplane, the others approach the objective unmolested. If an attempt is made to illuminate and fire at each airplane, dispersion results.

In the example given above, thirty-six airplanes deliver an attack within a period of fifty-five minutes. The droning of the engines of the airplanes approaching and leaving the objective at different altitudes should create a confusion of sound which should affect adversely the efficiency of the listening devices. With the muffling of engines and use of geared propellers, the drone of the incoming bombardment will be materially reduced.

This method of conducting a night attack is particularly important when hostile pursuit airplanes are cooperating with the searchlights. Once illuminated by several searchlights, it is most difficult, if not impossible, for a bombardment airplane to get out of the glare. Pursuit airplanes may approach the illuminated bombardment airplane with complete surprise. The defending machine gunners will not, ordinarily, see the attacking pursuit until the staccato sound of its machine guns are heard. If it attacks by long range fire, it may not even be seen. The approach is so rapid and direction so uncertain that the bombardment gunners will have extreme difficulty in meeting a pursuit attack. Thus, when night pursuit is operating all measures herein discussed and those to be treated under "special support by other classes of aviation," will be of utmost importance.

Special Support by Other Classes of Aviation

All bombardment operations receive special or general support by other classes of aviation. General support by observation, pursuit and attack aviation is considered to be as follows:

Observation: Reconnaissance missions performed particularly for the higher command, but which gains information of value to bombardment units in depicting particular bombardment targets.

Pursuit: Operations conducted over areas in which bombardment units, in addition to units of other classes of aviation, may operate.

Attack: Operations conducted against objectives which are not particularly bombardment targets but which will contribute to the mission of bombardment units. Such missions carried out simultaneously as to time with bombardment missions will cause diversion of the hostile pursuit. That is, the hostile pursuit has two or more formations to intercept and attack rather than one, thus lessening the chances that either will be intercepted. When the attack missions are conducted against hostile aircraft on the ground which might be employed against the bombardment units, general support is also performed.

The special support of bombardment aviation consists of those missions performed by the other classes of aviation to enhance primarily the success of the bombardment mission. Thus, observation units will be charged with securing photographs of specific bombardment objectives. Such photographs are of the utmost value to the bombardment commander, in planning the method by which the objective will be attacked; the direction of approach to the target; the division of the objective into particular targets for each subordinate bombardment unit. Upon completion of an attack, observation units will often be charged with reconnoitering the objective, securing photographs thereof and reporting the results of the bombardment attack. As study of the objective, prior to a bombardment attack, is of the utmost importance to the success of the mission, special support will be performed ordinarily by observation aviation in securing information and photographs of each bombardment objective.

In day operations the strength of the hostile pursuit in the area wherein the bombardment objective lies may be such as to make desirable the support of a bombardment formation by friendly pursuit. Friendly pursuit, in providing such support will rendezvous with the bombardment formation before it enters hostile territory and accompany it to and from the objective. When an objective lies beyond the radius of action of the friendly pursuit, the latter will accompany the bombardment formation as far as possible into hostile territory and then return to the front lines or to the pursuit airdromes. The pursuit unit which returns to its airdrome will reservice, clear its airdromes, and will meet the returning bombardment formation within the hostile lines at a predetermined point and accompany it to friendly territory, or to its airdromes. The primary mission of the supporting pursuit will be to engage hostile pursuit which attempts to attack the bombardment formation. The friendly pursuit remains always within supporting distance of the bombardment formation and by its action or presence prevents a coordinated attack against the formation. It is not expected that the supporting pursuit will be able to prevent hostile pursuit from making some attacks against the bombardment formation. However, it should be able to prohibit a coordinated attack. Against an uncoordinated attack a bombardment formation is particularly capable of taking care of itself.

Pursuit support is not required at night. Attacks by hostile pursuit against bombardment airplanes in flight are practically impossible unless the bombardment airplanes are illuminated by searchlights. Friendly pursuit cannot attack the hostile pursuit unless the latter is illuminated, which is unlikely.

Attack aviation will support a bombardment mission by neutralizing so far as is possible the antiaircraft artillery defending the bombardment objective. Fragmentation bombs may be employed against known antiaircraft gun emplacements. Smoke may be used against areas in which the antiaircraft batteries are likely to be. In supporting a low altitude operation by bombardment, attack can be used to clear the way of any balloon barrages established around the objective. The attack unit may lay a smoke screen over which the bombardment unit will fly, thus affording the latter a measure of protection from hostile machine gun and rifle fire. The support which attack aviation may provide for bombardment is of great assistance to bombardment in many night operations. In supporting a night bombardment mission, the attack airplanes will operate in the area surrounding the objective, throughout the period in which the bombardment attack is conducted. The attack airplanes will attempt to put out of action by use of either or both fragmentation bombs and machine guns, all searchlights which attempt to illuminate the incoming bombardment airplanes. The noise created by the low flying attack airplanes will contribute to the confusion of sound which should affect adversely the efficacy of the listening horns. By the use of parachute flares, the attack airplanes arriving shortly before the bombardment airplanes, may illuminate the target. By such support, attack aviation may make an important contribution to the success of the bombardment attack.

Conclusion

An attempt has been made to outline, in a most general way, those methods by which a bombardment attack may be driven home. There are many alternate methods which are believed practicable but which follow to an extent those above discussed. How efficacious the tactics may prove to be can only be discovered in war. It is believed, however, that proper application of these tactics will lead to successful bombardment operations.

By no means may it be assumed that bombardment units, applying these or any other tactics, can avoid casualties. A certain loss in men and material is a price which must be paid for success. This is true for any military force. The results obtained will be determining factor as to whether or not the price paid is too high.

It must be remembered by those responsible for the defense against bombardment operations, that a bombardment unit will not be stopped by the presence of a strong defense or a mere show of force. It is generally conceded, by those who are competent to judge, that an air attack well launched is most difficult to stop. The bursting of antiaircraft shells or the presence of a hostile pursuit force will not prevent a determined bombardment commander from accomplishing his mission. To stop a bombardment attack, the bombardment airplanes must be shot out of the sky.

In the final analysis, the most efficacious method of stopping a bombardment attack is to destroy the bombardment airplanes before they take the air. As a bombardment unit will be upon its airdrome *at least* sixteen out of every twenty-four hours, the *best defense* would appear to be *an offensive* against the bombardment airdrome.

Bombardment Aviation—Bulwark of National Defense

LIEUT. K. N. WALKER
Air Corps Tactical School, Maxwell Field, Alabama

BOMBARDMENT aviation has confined wars to continents! Successful invasion of the United States by an armed enemy is impossible in the face of an adequate force of bombardment airplanes manned by trained personnel. These statements contain a fundamental proposition upon which our National Defense may be predicated.

Bombardment aviation is the principal component of an air force, which consists also of attack, observation and pursuit aviation. By reason of the great striking power of the bombardment airplane, bombardment aviation is the basic arm—the backbone of any air force. The operations of this basic component of an air force will, in any major war, be paramount. In organizing an air force, we must first determine the amount of bombardment aviation required, and then build a balanced force by assigning to it the necessary attack, observation and pursuit aviation to insure that the bombardment attack will be driven home.

It is obvious that the air force is a team, with bombardment carrying the ball. When we speak of bombardment aviation or of an air force, we are thinking in terms of this team. Whenever we speak in terms of "air force" we are thinking of bombardment aviation. Tactical and technical developments will influence and change the ratio of the several supporting members of this air force team with reference to the bombardment component. But whatever the ratio, we must have enough bombardment airplanes with which to deliver that destructive blow upon which the air force depends for success.

Before considering those situations wherein bombardment may play a leading role in national defense, air force

---

Walker, Kenneth N., "Bombardment Aviation: Bulwark of National Defense," *U.S. Air Services*, August 1933, 15–19.

employment may be outlined. An air force is employed properly against those establishments necessary to the enemy's prosecution of war.

The fact that logistics (the art of supply and transportation) is the Achilles Heel of modern warfare is fast becoming recognized. Destruction of the means by which any armed force is supplied, moved and made ready to operate, eliminates that force as threat. There will be certain vital enemy objectives, which if destroyed will contribute most to the success of the combined arms of the nation. Those objectives may be the hostile naval force which insures the maintenance of the enemy sea lines of communication; convoyed merchantmen transporting air and land forces to bases from which a blow can be struck against us; dock facilities, supply establishments and lines of communications on land necessary for the unloading, supplying and moving of hostile forces; operating airdromes of hostile air units. Successful air operations against such vital objectives will hamstring an enemy.

NO ENEMY WOULD CONSIDER launching an invasion against the United States if he were convinced that we were in possession of a bombardment force capable of destroying such targets. An adequate air force built around bombardment aviation can effect such destruction. By its ability to conduct operations against such objectives as outlined, an air force insures against invasion. An air force does not wait for an enemy to don leisurely his mailed gloves and prepare for action—it strikes at his vulnerable points while he is in the process of preparation.

If the United States is ever attacked, our enemy will be represented by a coalition of foreign powers. No one state will ever attempt an invasion without some hope of success. No one nation, unaided by others, could hope for successful conclusion of hostilities. Hence, we must plan to defend against a coalition which has a possibility of concentrating superior forces against us. We must prepare to meet the worst, rather than the best situation, for it will probably be the worst that we must meet.

A FOREIGN COALITION of major powers would be capable of bringing superior naval forces against our own fleet. If such superiority should result in the defeat or destruction of our fleet, the sea lines of communication would then be open to the enemy. Troop ships, no more opposed than were our own during the World War, could conceivably bring superior ground and air forces to bases on territory adjacent to the United States, and keep them supplied.

Disregarding air action for the moment, it is clear that under such circumstances we might be invaded by superior ground forces before we could mobilize our own ground troops in sufficient numbers to meet them. The situation would be different than in 1917, when the armies and navies of several nations held the barrier and gave us time in which to prepare.

We must be ready to meet successfully an enemy, intent upon invading this country—an enemy who first by massing superior naval forces against us can assure the safety of his lines of communications for the transport of troops and supplies necessary to invasion. We must be prepared to meet such a situation and the answer lies in an adequate air force, organized around bombardment aviation, that can operate independently or in support of both our naval and ground forces—an air force that can meet an attack directed against any of our frontiers—an air force operating upon interior lines that can be concentrated within a short period to meet the major threat.

AN ENEMY, intent upon successful invasion, must attempt initially to dispose of our own Navy. With superiority in numbers, the enemy's engagement of our fleet might end in disaster to the latter—*unless* by some means we would support our fleet. It is submitted that an adequate air force, operating from land bases, can, within its radius of action, provide that support to our own fleet which will result in so decimating the opposing fleet that our own can meet it on equal or superior terms.

The potentialities of a determined air attack against a fleet are such that a question arises as to whether a naval commander will ever subject his force to the hammer blows of demolition bombs from an air force on the mere chance that fighter aircraft and antiaircraft guns might prevent it. Our Navy, by operating within the radius of action of our land based air force, could secure its support. Such action does not appear to be illogical—history records that no major naval engagement has ever been fought so far from shore that it would have been outside the reach of land based aircraft.

If the hostile fleets were imbued with the desire or necessity of destroying our Navy, they would then be forced to accept battle where the latter desired it. Our Navy could well withdraw and refrain from commitment to action until our air force had so "whittled down" the enemy forces that they were no more than equal and possibly inferior. The results of the air attacks would probably be such that the hostile navy would seek to escape further attacks from the air and an engagement with our fleet.

THERE is every reason to believe that attacks by an adequate air force against a hostile fleet would be crowned with success. Demolition bombs can be dropped with accuracy sufficient to hit naval targets. Such hits against carriers or battleships will either destroy, damage seriously, or at least put out of commission those vessels and prohibit them from taking part in an imminent naval engagement. The problem of attacking a hostile fleet is not simple, nor is it difficult—no more so than an engagement between opposing naval forces or opposing ground forces. Opposition must be expected from carrier based fighter aircraft and from the antiaircraft guns aboard ship. Nevertheless, study leads to the belief that the casualties in airplanes would not be sufficiently heavy to render ineffective the air attack. As a result of recent exercises conducted by the British Home and Mediterranean Fleets, where aircraft were employed against naval forces, the naval correspondent of the British publication, *The Fighting Forces*, has this to say:

> Actually, when we have got over the first initial thrill of seeing swarms of aircraft swooping down on the target like golden eagles on their prey, we should ask ourselves: What has the demonstration proved? The answer is that, *provided there is no opposition*, reconnaissance planes can locate a fleet and pass on the necessary information; bombers can drop bombs accurately from a great height on a moving target and torpedoplanes can register a reasonable proportion of hits with their weapons. But only provided there is no opposition.

HERE is an admission that success may attend an air attack against a fleet—provided there is no opposition. But we should ask ourselves another question: Can a fleet be provided with sufficient air support to meet and defeat a land based air force? Can pursuit or fighter aircraft, under the difficulties imposed in operating from the vulnerable carrier, be massed at the proper time and place and in sufficient numbers to meet and defeat the well planned, coordinated attack of a land based air force?

We think not. Military air men of all nations agree that a determined air attack, once launched, is most difficult, if not impossible to stop when directed against land objectives. How much more pertinent is this conclusion when the objective is a naval force. In developing air tactics with which to meet a hostile fleet, all probable opposition must be considered carefully.

Let us consider for a moment the tactics that might be employed by an air force operating against a battle fleet. The mission of the air force and of the bombardment component is to destroy those vessels which are the greatest menace. The operations of other classes of aviation are coordinated so as to drive home the bombardment attack.

Observation planes locate and report the dispositions of the fleet. Attack airplanes precede the bombers by a few short minutes and neutralize ship-board antiaircraft batteries with smoke and bombs and, if the situation demands, destroy the flight decks of the carriers with the smaller demolition bombs. Pursuit (or fighter) aircraft (single, bi or multiseat) may support both the attack and bombardment operations by engaging and preventing the carrier based fighters from launching well coordinated attacks against the bombardment and attack formations. The bombardment tactics are developed along lines which will afford the greatest security from pursuit and antiaircraft opposition, while at the same time insuring the placing of bombs on the selected targets.

BOMBARDMENT AVIATION may play a vital part in the defense against invasion in a situation wherein an enemy secures air bases on territory adjacent to the United States. From such bases, air support might be given to forced landings of troops on our shores, or air operations might be directed against our vital industrial establishments. Such action might be attempted, with the hostile navy refusing to engage our fleet within radius of action of our air force and operating outside thereof primarily for the purpose of protecting the lines of communications over which air and land forces and supplies must be transported.

However, to be within reach of vital targets within our territory, the hostile bases would of necessity be within radius of action of our air force. We would not await attacks and attempt only to meet them from the air with pursuit aviation and from the ground with antiaircraft artillery. We would seek out and destroy those bases. To avoid the possibility of similar enemy air action against our own base airdromes, the latter could be located outside the radius of action of the enemy air units.

Then by our use of advanced airdromes, where supplies of fuel only would be kept, we could extend the radius of action of our aircraft so as to include the area in which the enemy is establishing his bases, and operate against them, while at the same time similar action would be impossible for him. Thus we could secure our own bases while making untenable the hostile bases.

With an adequate air force, which could destroy the hostile airplanes on the ground and the supplies and facilities necessary to their operations, we would prohibit air operations against us.

While an enemy, intent upon the above type of operations, might prepare to bring a superior number of airplanes against us, it is rather improbable that all of the air units could be transported and prepared for action simultaneously. We would have the opportunity of defeating in detail the potentially superior air force.

This would only be possible if we had an adequate air force, trained and organized to meet the threat when it appears. Should we be forced to wait from six months to a year after declaration of hostilities to place in the field this adequate air force, it might be too late.

We must appreciate the fact that an air force, capable of operations against those land bases from which hostile air attacks may be launched against our vital centers or in support of forced landings on our shores, can prevent such hostile action and we must conclude that such an air force is well worth while.

A THIRD SITUATION may be considered. Should an enemy succeed in landing troops in adjacent territory with the mission of invading our country with land forces, our ground troops would be concentrated as quickly as possible to meet that threat. The enemy would not contemplate or attempt such action without air support, however. The operations of our air force under such circumstances would certainly determine the tide of the conflict.

Initially, our air force would provide security for our troop concentrations and for the friendly logistical establishments by attacks against the hostile air force. Coincident with such operations, our air units would operate against the rail and vehicular lines of communications necessary for the movement of troops and supplies and against the supply establishments required for their maintenance.

Successful air operations against such objectives could alone prevent an invasion contemplated along the above lines. No land force will attempt forward movement in the face of destruction of its lines of communications and supplies, and the latter afford two of the most profitable targets for air attack. Success would attend our air force operations, provided that air force was adequate.

AS POINTED OUT PREVIOUSLY, a possible invader must consist of a coalition of foreign powers—not one single nation. It is doubtful that one nation alone could bring to our shores both naval and land forces superior to those which could be assembled by the United States. If we do not consider as a possibility an offensive war ever being waged against us, then the maintenance of forces for the defense of this country is monetary waste. If, on the other hand, we do consider that possibility, but sustain a force insufficient to the purpose, can we hope to hold the levee when the flood of war bursts upon us?

For economic reasons, it is impossible for us in time of peace to build a navy sufficient in numbers to meet on equal terms the fleets of a foreign coalition of major powers. It is against our principles to maintain a standing army comparable to that of any one of the great powers, let alone a coalition.

What to do? It is proposed that it is possible for us, both economically and in accordance with our principles, to build an adequate force of bombardment aviation, together with the necessary supporting classes of aviation, to meet and defeat, by independent action and in support of our Navy and Army, the forces of a foreign coalition which might be brought against us. We can and will do this when the importance of an air force to our national security is appreciated fully—not in terms, for example, of a mere adjunct to our ground forces, such as cavalry or field artillery and designed *only* to further the infantry mission—but as a force with a distinct mission, of importance co-equal to that of the Army and the Navy.

We must learn not to use a sledge hammer to drive tacks.

IT MAY BE PERTINENT to emphasize, at this point, that none of these or similar situations, discussed above in theory, would ever develop into actuality with the United States in possession of this adequate air force. An enemy will never initiate an attempt at invasion if it is foredoomed to failure upon arrival of his forces within radius of action of a powerful air force of bombers.

Much has been written concerning the means by which a bombardment attack may be nullified. Statements have been made to the effect that tests and experience during the past few years have shown that a bombardment airplane can exist no longer than twelve seconds in the face of antiaircraft artillery. Such statements are entirely fallacious and unsupported by facts, tests, or experience. Study and analysis confirm the opinion that while antiaircraft fire is a menace, will interfere with leisurely bombing, will cause casualties, it will not stop a bombardment attack conducted in accordance with tactical principles now in practice, nor prevent the bombardment force from accomplishing its mission.

Hostile pursuit aviation is the greatest menace to the success of the bombardment mission. But first the hostile pursuit must intercept the bombers, in sufficient time before the bombardment objective is reached, to launch an effective attack which may interfere with or prevent the accomplishment of the mission.

WITH BOMBARDMENT AIRPLANES rapidly approaching the speeds of foreign pursuit planes, the problem of interception is becoming more and more difficult. After interception is made, the big task of shooting down the bombers must be undertaken by pursuit. A bombardment formation is inherently capable of delivering superior fire against the attackers and, in the final analysis, the result will hinge on fire power.

Should the pursuit threat ever become so serious that a bombardment formation must expect serious losses, accompanying bi or multi-seater pursuit will provide the necessary support and augment the fire power of the bombardment formation, and thus contribute to the success of the bombardment attack. Study of war records offers ample proof that with proper defensive tactics bombardment aviation may be expected to perform its mission in the face of pursuit opposition. The following quotations are illuminating:

> The big lesson learned from the many combats with the Germans' ace squadron which had been moved to the sector was that big tight formations were necessary to successful bombing operations. The heavy losses were due to small formations of three and four airplanes being wiped out. A large formation with a tight rear line is almost invulnerable.—*Extracted from records of the American 96th Bombardment Squadron.*

> Our bombing formations had constantly to fight their way to their objectives. Enemy aircraft were encountered in large formations on every raid. They showed great reluctance to close with the DH4's when in good formation and the aeroplanes we lost were all brought down by long distance shooting.—*Records of the British No. 57 Squadron.*

> Our bombardment groups were attacked every time they took the air and statistics prove that—a few cases being excepted—our combats with German aviation always resulted in losses to them at least equal, but more often superior to ours. *Commandant De La Morlais of the French Air Division.*

COMMANDANT DE LA MORLAIS goes on to point out that the five day bombardment groups, consisting of about 225 airplanes and organized into the French Air Division, dropped during an eight months' period from March to October 1918, 1,665 tons of bombs. During this period that force shot down 145 German pursuit airplanes and lost 66 airplanes within German lines. Analysis of the data available shows that the losses approximated only 1.3% of the airplanes per mission.

War records exemplify the point that a bombardment formation properly flown and resorting to proper tactics did not suffer losses out of proportion to the results gained. It is well to note that, during the war, the performance of the bombardment airplane was always far behind that of the pursuit airplane. Today, when we consider the exceptional performance of our latest bombardment airplanes, compared with the pursuit airplanes of foreign nations; when we realize that bombardment tactics have developed to a high state of efficiency; it is obvious that, while pursuit will cause casualties, those casualties will rarely be sufficient to stop a bombardment unit from performance of its mission.

The only reliable means of preventing an air attack is to stop it before it gets started. Destruction of airplanes upon their airdromes (where they will be found 75% of the time); destruction of the supplies and facilities necessary to their operation; and destruction of the means by which supplies are moved to the operating airdromes, are the only sure methods of preventing air operations. An enemy intent upon invading this country cannot hope for success unless he can place a superior force in the air against us and can maintain that superiority by the above means. With an adequate air force, built around bombardment aviation, trained and ready for immediate operations, it is inconceivable that superior numbers could be assembled in time to "beat us to the punch."

THE BOMBARDMENT AIRPLANE is recognized as the acme of offensive power.

"The modern bomber is the most deadly and most effective weapon of modern warfare."—*Maj. Gen. Mason M. Patrick.*

"The bombing airplanes are the outstanding offensive weapon from the operative viewpoint in the conduct of war which creates terror in the minds of populations."—*Colonel Strecius, of Germany.*

The possession of a large force of bombardment airplanes by a European nation might be considered a constant threat to the peace and security of a neighboring nation. The development of the bombardment airplane and of bombardment aviation has placed in the hands of a belligerent a weapon which is capable of reaching out and destroying an enemy's means for the prosecution of war, immediately upon declaration thereof.

Such a force in the hands of an unscrupulous state could easily cause the grave concern of other nations and result in a race for superiority in bombardment airplanes. Any nation, inferior in bombardment aircraft, is at a grave disadvantage. The concern which all European nations feel over the possibilities of aerial bombardment is evidenced particularly by the steps which are being taken with respect to the means of passive defense against this type of warfare.

And so, when nations meet to propose limitations of armament, it is consistent that first consideration be given to those weapons which are purely offensive in character. And in the list of weapons in that category, the bombardment airplane may be expected to lead the list. From the point of view of a European nation, the elimination of the bombardment airplane removes a means by which instant and effective attacks may be launched against its vitals by a hostile state.

IT IS ASSUMED that by elimination of offensive weapons offensive warfare is less likely. However, when we consider the bombardment airplane from the viewpoint of the United States we discover a seeming paradox. *The bombardment airplane in the hands of the United States is a purely defensive weapon.* It is impractical to construct bombardment airplanes capable of spanning the ocean and returning, and at the same time capable of releasing destructive loads of demolition bombs on the nerve centers of some European nation.

Not only is it impractical, but it is undesirable. We require bombardment airplanes capable of radii of action well within 1,000 miles. A bomber with a radius of action of 500 miles at the present maximum speeds will fit our needs. Therefore, we cannot reach nor can we threaten the peace of a foreign nation that might be a possible enemy with a force of bombardment airplanes.

But, with an adequate force of bombers, we *can* insure the defense of the United States. Our desire for a strong force of such airplanes can in no way be interpreted by another nation as an indication of our desire for war. It *can* be interpreted as a desire for peace. The possession of such a force is evidence that not only do we insist upon the maintenance of our integrity, but that we are capable of maintaining it.

UNFORTUNATELY, we have no numerous nor well established precedents by which we may support our assertions. The air operations of the World War, although establishing their potentialities, leave much to be desired. We can only stand upon the tests of logic, of reason, and of peacetime maneuvers and exercises. While we regret the lack of precedent, we are not checkmated by its absence. So long as National Defense is of importance, we must prepare to take full advantage of those weapons which will contribute most to the protection of our country. The possibility of ultra-conservatism due to lack of precedent was appreciated by Marshal Foch when he said:

> The military mind always imagines that the next war will be on the same lines as the last. This has never been the case and never will be. One of the great factors in the next war will obviously be aircraft. The potentialities of aircraft attack on a large scale are almost incalculable, but it is clear that such attack, owing to its crushing moral effect on a nation, may impress public opinion to the point of disarming the government and thus become decisive.

In this country, the steps which have been taken toward development of air tactics and toward technical development of airplanes prove conclusively that we are awake to the potentialities of air defense. We have laid the foundation for that defense. We can conclude that bombardment aviation has confined wars to continents and, so far as this country is concerned, it offers a new and most effective means of security.

[Author's Note: The statements and opinions made herein are those of the writer and do not represent the official views of the Air Corps.—K. N. W.]

Bibliography

Books

The Air Force, text, April 1930, 248.101-1 USAFHRC.

Anders, Curtis. *Fighting Airmen*. New York: Putnam's, 1966.

Assistant Chief of Staff, Intelligence. *Air Action in the Papuan Campaign, 21 July 1942 to 23 January 1943*. USAF Historical Study 17, 1944. Maxwell AFB, Ala., Air University Library Documents Section, M-U 27218, no. 17.

Barker, J. D. *History of the Air Corps Tactical School*. Maxwell AFB, Ala.: United States Air Force Historical Research Center (USAFHRC), 245.01-B.

Byrd, Martha. *Chennault: Giving Wings to the Tiger*. Tuscaloosa, Alabama: University of Alabama Press, 1987.

Chennault, Claire L. *Way of a Fighter*. New York: G. P. Putnam's Sons, 1949.

Copp, DeWitt S. *A Few Great Captains: The Men and Events That Shaped the Development of U.S. Air Power*. Garden City, N.Y.: Doubleday, 1980.

________. *Forged in Fire: Strategy and Decisions in the Air War Over Europe, 1940-45*. Garden City, N.Y.: Doubleday, 1982.

Craven, Wesley Frank and James Lea Cate, eds. *The Army Air Forces in World War II*, vol. 1, *Plans and Early Operations January 1939 to August 1942*. Chicago: University of Chicago Press, 1948.

________. *The Army Air Forces in World War II*, vol. 4, *The Pacific: Gualaldanal to Siapan August 1942 to July 1944*. Chicago: University of Chicago Press, 1958.

DuPre, Flint O. *U.S. Air Force Biographical Dictionary*. New York: Franklin Watts, 1965.

Fabyanic, Thomas A. "A Critique of United States Air War Planning, 1941–1944." PhD diss., Saint Louis University, 1973.

Finney, Robert T. *History of the Air Corps Tactical School, 1920–1940.* Washington, D.C.: Center for Air Force History, 1992. Originally published: Research Studies Institute, USAF Historical Division, Air University, 1955; USAF Historical Study 100, Maxwell Air Force Base: Research Studies Institute, 1955.

Foulois, Benjamin D., with C. V. Glines. *From the Wright Brothers to the Astronauts: The Memoirs of Benjamin D. Foulois.* New York: McGraw-Hill, 1968.

Futrell, Robert Frank. *Ideas, Concepts, Doctrine: A History of Basic Thinking in the United States Air Force, 1907–1964.* Maxwell AFB, Ala.: Air University, Aerospace Studies Institute (1971; reprint, Maxwell AFB, Ala.: Air University Press, 1989).

Gaston, James C. *Planning the American Air War: Four Men and Nine Days in 1941, an Inside Narrative.* Washington, D.C.: National Defense University Press, 1982.

Gillison, Douglas. *Royal Australian Air Force, 1939–1942.* Canberra: Australian War Memorial, 1962.

Goddard, George W., with DeWitt S. Copp. *Overview: A Life-Long Adventure in Aerial Photography.* Garden City, N.Y.: Doubleday, 1969.

Greer, Thomas H. *The Development of Air Doctrine in the Air Arm, 1917–41.* USAF Historical Study 89. Maxwell AFB, Ala.: Air University, USAF Historical Division, Research Studies Institute (1955; reprint, Washington, D.C. , Office of Air Force History 1985).

Hansell, Haywood S., Jr. *The Air Plan that Defeated Hitler.* Atlanta: Higgins-McArthur/Longino and Porter, 1972.

________. "Harold L. George: Apostle of Air Power." *Makers of the United States Air Force.* Edited by John L. Frisbee. Washington, D.C.: Office of Air Force History, 1987.

Higham, Robin. *Airpower: A Concise History.* New York: St. Martin's Press, 1972.

Hurley, Alfred Francis. *Billy Mitchell: Crusader for Air Power.* New York: Franklin Watts, 1964.

Kenney, George C. *General Kenney Reports: A Personal History of the Pacific War, With Maps By the Author.* New York:

Duell, Sloan, and Pearce (1949; new imprint, Washington, D.C.: Office of Air Force History, 1987).

Kirkpatrick, Charles E. *An Unknown Future and A Doubtful Present: Writing the Victory Plan of 1941*. Washington, D.C.: Center of Military History, United States Army, 1990.

Lambert, John W. *The Pineapple Air Force: Pearl Harbor to Tokyo*. St. Paul, Minn.: Phalanx Publishing Co., Ltd., 1990.

McAulay, Lex. *Battle of the Bismarck Sea*. New York: St. Martin's Press, 1991.

Meilinger, Phillip S. *American Airpower Biography*. Maxwell AFB, Ala.: School of Advanced Airpower Studies. Air University Press, July 1995.

Millis, Walter. *Arms and Men: A Study in American Military History*. New York: Putnam's, 1956.

Purtee, Edward O. *History of the Army Air Service, 1907–1926*. Wright-Patterson AFB, Ohio: Historical Office, Air Materiel Command, 1948. Maxwell AFB, Ala.: Air University Library, call #358.09 A2982h, for library use only.

Schaffer, Ronald. *Wings of Judgment: American Bombing in World War II*. New York: Oxford University Press, 1985.

Sherman, William C. *Air Warfare*. New York: The Ronald Press Company, 1926.

Sherry, Michael S. *The Rise of American Air Power: The Creation of Armageddon*. New Haven: Yale University Press, 1987.

Training Directive, 1 July 1938–30 June 1939. GP-5-SU-DI (BOMB) USAFHRC, Maxwell AFB, Ala.

________. GP-5-HI (BOMB), 1918–1945, USAFHRC.

United States Army Air Corps. *Bombardment Aviation*. Air Corps Tactical School, Army Extension Courses Special Text no. 199; 1934 and 1938. USAFHRC Number 248.1011-199.

________. *Bombardment Aviation*. Prepared under direction of the Chief of Air Corps. Washington, D. C.: US Government Printing Office, 1938. Maxwell AFB, Air University Library Call Number 623.74 U58b.

Webster's American Military Biographies. Springfield, Mass.: G. & C. Merriam Co., 1978.

Williams, Kathleen. *Army Air Forces in the War Against Japan 1941–1942.* USAF Historical Study 34, 1945. Maxwell AFB, Ala.: Air University Library Documents Section, M-U 27218, no. 34.

Articles

Aerospace Historian, 25–26.

Birdsall, Steve. "Target: Rabaul!" *Air Force Magazine* 58 (September 1975): 108–13.

The Denver Post, 23 December 1937.

Frisbee, John L. "Skip-Bombing Pioneer." *Air Force Magazine* 73, no. 12 (December 1990): 87.

Gage, Philip S., Jr. "Fliers and Medals of Honor." *Aerospace Historian* 30, no. 4 (December 1983): 254–60.

Hansell, Haywood S. "Brig. Gen. Kenneth N. Walker, Prophet of Strategic Airpower." *Air Force Magazine*. November 1978, 92–94.

________. "With His Boots On." *Time*, vol. 41, 5 April 1943, 73.

"Pilots Show Speed of Bombing Planes." *New York Times*, 11 September 1926, 17.

"Pacific Mission." *Life* 14, no. 6 (8 February 1943), 97.

Sheehan, Susan. "A Missing Plane." *The New Yorker* 62 (12 May 1986): 46–48; 62 (19 May 1986): 43–44; 62 (26 May 1986): 39–40.

Taylor, Joe Gray. "They Taught Tactics." *Aerospace Historian* (Summer 1966), 67–68,

Walker, Kenneth N. "Bombardment Aviation: Bulwark of National Defense." *US Air Services*, August 1933, 15–19.

________. "Driving Home the Bombardment Attack." *Coast Artillery Journal* 73, no. 4 (October 1930): 328–40.

Wilson, Donald. "Origin of A Theory for Air Strategy." *Aerospace Historian* 18, no. 1 (March 1971): 19.

Oral Histories

Davidson, Howard C., Maj Gen. Oral History interview with Hugh N. Ahmann and Thomas A. Sturm. 5–8 December 1974, K239.0512-920, USAFHRC.

Kuter, Laurence S., Gen. Oral History interview with Hugh N. Ahmann and Thomas A. Sturm. 30 September–3 October 1974, K239.0512-810, USAFHRC.

Partridge, Earl E. Oral History interview with Frank E. Merritt. K239.0512-1112.

Peabody, Hume. Oral History interview with Hugh Ahmann and Thomas A. Sturm. 30 September–3 October 1974, K239.0512-810, USAFHRC.

Saville, Gordon P., Maj Gen. Oral History interview with Thomas A. Sturm. 26–29 March 1973, K239.0512-1322, USAFHRC.

Wilson, Donald. Oral History interviews with Hugh N. Ahmann. K239.0512-878 and K239.0512-810, USAFHRC.

Letters, Notes, and other Correspondence

Accident Report. 200.3912-1, USAFHRC.

Adjutant General. Letter to CGAFF, 14 January 1943. K141.2421, USAFHRC.

Aldworth, Richard. Letter to Lauchlin Currie, 10 September 1941.

Andrews, Bill. Letter to Douglas Walker.

AWPD-1 Scrapbook. 145.82-1, USAFHRC.

CGAFF. Letter to Adjutant General, 29 January 1943. K141.2421, USAFHRC.

Lauchlin Currie Papers, Box 2, Hoover Institution.

Diary of Henry H. Arnold, 25 September 1942. 168.65-42, USAFHRC.

Diary of Charles L. Hodge.

Diary of William Travis, 5–9 January 1943.

Diary of William Whitacre, 5 January 1943.

Federal Aviations Commission Testimony. 248.121-3, USAFHRC.

Federal Aviations Commission Report. 248.211-53, USAFHRC.

Hansell, Haywood S., Jr. Letter to Marguerite Walker, 16 July 1986.

Hassemer, David W. San Antonio, Tex. Letter to author, 21 January 1991.

Holloway, Bruce K. Orlando, Fla. Letter to author, 19 July 1990.

Hoy, Bruce. Aspley, Australia. Letters to Gene Monihan, 23 May 1988 and 3 June 1988.

Holguin, Jose L. Letter to Douglas Walker, 19 September 1982.

Hughey, Robert J. Camas, Wash. Letter to author, 21 January 1991.

Hurley, Alfred. Letter to author, 18 February 1991.

Kenney, George C. Letter to Henry H. Arnold, 1 January 1943. 706.168-2, USAFHRC.

Kenney Papers, vol. 1, Office of Air Force History, 5 August 1942; 5, 8, 12, 15, and 23 October 1942; 18 November 1942; and 10 December 1942.

Kuter, Laurence. Unpublished autobiographical manuscript. USAF Academy Library, Special Collections.

McMurria, James A. Letter to Gene Monihan, 30 May 1991.

Memorandum. 21 February 1941, Maj K. N. Walker to Gen Carl A. Spaatz.

Memorandum. 11 August 1941, Maj Haywood S. Hansell Jr., to Gen Henry H. Arnold.

Memorandum. 12 September 1941, to General Echols. 145.82-1, USAFHRC.

Memorandum. 15 September 1941, Lt Col Harold George to Chief of the Air Staff, subject: AWPD-1. 145.82-1, USAFHRC.

Memorandum. 25 September 1941. 145.82-1, USAFHRC.

Monihan, Gene. Letter to Bruce Hoy, 24 June 1988.

________. Letter to James A. McMurria, 11 August 1991.

Pearson, Bob. Letter to Douglas Walker, 1 September 1964.

Robinson, Gerald G. Riverside, Calif. Letter to author, 18 January 1991.

Sanxter, Donald L. Letter to Gene Monihan, 24 February 1991.

Shelton, Douglas C. Little Rock, Ark. Letter to Douglas P. Walker, 18 January 1991.

Transcript. Bill Henry's "By the Way" broadcast, 25 March 1943.

Travis, William W. Birmingham, Ala. Letter to author, 5 August 1991.

Walker, Kenneth N. Copy of Will, 19 June 1942. AG210.311.

________. Letter to Major Weaver, 5 February 1932.
________. Letter to Carl A. Spaatz, 3 September 1932.
________. Letter to Ken, 1 March 1942.
________. Letter to Ken, 1 May 1942.
________. Letter to sons, 15 June 1942.
________. Letter to Ken and Doug, 21 June 1942.
________. Letter to sons, 11 August 1942.
________. Miscellaneous notes to Willoughby, 22 September 1942. HRS(I) A-104, NA.
Walker Medal of Honor Scrapbook.
Walker, Kenneth N., Jr. Letter to author, 8 March 1991.
Walker, Marguerite. Letter to her parents.
Whitehead, Enis. Letter to Donald Wilson, 27 March 1943.

Interviews

Andrews, Mamie Lee (Mrs. W. K.). Chatham, Mass.
Barton, Mary Lee (Mrs. Charles). Interview with author, 16 October 1990.
Hassemer, David W. San Antonio, Tex.
Hickey, Lawrence. 43d Bomb Group Historian. Interview with author, 28 November 1990.
Hodge, Charles L. Interview with author, 16 October 1990.
Hodge, Mary (Mrs. C.L.). Arlington, Va. Interview with author, 16 October 1990.
Holloway, Bruce K. Orlando, Fla.
Knerr, Hugh T. Interview, 2 October 1971, from Walker Medal of Honor Scrapbook.
Kuter, Ethel (Mrs. L.S.). San Antonio, Tex. Telephone interview with author, 1991.
Monihan, Gene. Haymarket, Va. Telephone interviews with author, 8 August 1990 and 15 October 1990.
Norton, Patrick H. Monroe, Mich. Telephone interview with author, August 1991.
Pearson, Scott. Denver, Colo. Telephone interview with author, 7 February 1991.
Perakos, John. Southing, Conn. Telephone interview with author, August 1991.
Shelton, Douglas C. Little Rock, Ark.

Travis, William W. Birmingham, Ala. Telephone interview with author, 4 August 1991.

Walker, Douglas P. New York, N.Y. Interview with author, 19 January 1992.

Walker, Kenneth N., Jr. Berkeley, Ontario.

Walker, Marguerite. Videotaped interview.

Whitacre, William L. Panama City, Fla.

Miscellaneous

ACTS, Bombardment Textbooks 1924–25, 1926, 1930, February 1931, October 1933, 30 July 1935, and 1 January 1938. 248.101-9, USAFHRC.

ACTS textbook, *The Air Force,* April 1930. 248.101-1, USAFHRC.

Adjutant General (AG)-cl.file 676.9A. Aircraft Warning Service (Cite partial copy in Walker's Medal of Honor Scrapbook).

AG-704 Missing Report, 5 January 1944. SR&D 299, RG407 NA.

AG-704 Missing Report, 12 December 1945.

AG-704 Missing Report Board Proceedings 2352, 20 July 1949. NA.

Air Service News Letter.

American Airpower in World War II. Haywood S. Hansell Jr. manuscript. K112.3-2, USAFHRC.

Associated Press. 21 September 1942.

AWPD, Plan for the Employment of Bombardment Aviation in Defense of Oahu, 14 August 1941. 145.811-25, USAFHRC.

Bissell, Clayton. Brief History of the Air Corps and Its Late Developments. Call no. 358.4073, 1 January 1927. US82b, Air University Library, Maxwell AFB, Ala.: USAFHRC.

Biographies, awards and decorations, and related public relations material. 13 January 1943, K141.2421, USAFHRC.

Fifth Bomber Command, vol. 2, 723.01, sec. III.

GP-18-HI, newspaper clippings. 22 March 1940, USAFHRC.

Hansell, Haywood S., Jr. "Strategic Air War against Japan." (Unpublished manuscript). Maxwell AFB, Ala.: Air War College, Airpower Research Institute, 1980.

Radio C-501, 16 September 1942, HRS(I) A-966, NA.

Radio 547-C, 21 September 1942.

Report of Missing Personnel Investigation Unit - 706.288 USAFHRC.

HRS(I)-1555, 1603, 2113, NA.

Walker's 201 File.

Index

11th Bombardment Squadron: 22, 29, 51, 126–127
18th Group: 57
18th Pursuit Group: 54–55, 57, 59
19th Bomb Group: 88, 90, 93, 96–97, 103
1st Transportation Group: 23
22d (M) with B-26s: 90, 98
28th Bombardment Squadron: 13–14
2d Bomb Group: 22, 50, 65
321st Squadron: 115
35th (P-40s) at Port Moresby: 90
38th (M): 90
3d Bombardment Group (L): 90
43d Bomb Group: 101, 103, 113
43d recipient of the Congressional Medal of Honor: 121
49th (P-40s) fighter groups: 89
59th Service Squadron: 21
5th Bomb Group: 53
5th Bomber command: 101
604th Quartermaster Graves Registration Company: 126
63d Squadron: 98, 100–101
64th Squadron: 113
7th Bomb Group: 50–52
8th (P-39s): 89
90th: 93, 103
9th Bombardment Squadron: 51
A-20: 90, 94
A-24: 90
AAF: 64–65
ABC-1: 66–68
aerial doctrine: 38
air combat groups: 22
Air Corps: xii, xix–xx, xxvii, 24–26, 29, 30, 33, 35–38, 43–45, 47–48, 50, 57, 63–64, 66, 71–72
  Act: 21
  Board: 70, 72
  Capability: 67
  Doctrine: 43

Materiel Division: 76, 91
Plans Division: 63
priorities: 72
Tactical School: xi, xv, xxii, xxv, 23–25, 29, 31–32, 35–36, 45, 50, 52, 57, 63, 67–68, 72–74, 91, 113
tradition: 57
Air Force: 24, 26–27, 29, 36, 43, 46, 68, 96, 117, 121
Combat Command: 64
Air Intelligence Section: 13
Air mission: 46
Air Service: xix, xxii, 4, 6, 10, 12, 15, 21–22, 23
Flying School: 4, 6
Newsletter: 14
Air Staff: 63
air strategy: 44
Air Transport Command: 23
Air War Plans: 76
Division (AWPD): xii, 58, 65
air's doctrinal evolution: 67
Aircraft Required for Control of the Seas: 69
airpower: xi, xx–xxi, 21–23, 28–29, 32, 38, 57, 64–65, 72, 74, 76, 94, 112
Alaskan Defense Command: 77
Aldworth, Richard: 63
Allied: 66, 87, 92, 102, 109, 125
air forces: 109
cooperation and strategy: 66
engineers: 92
ground forces: 102
Pacific strategy: 87
prisoners of war: 125
success: 87
Allies: xii–xiii, 64, 87, 92–93, 96, 107
Inn: 63
American : 33, 72–73
Volunteer Group: 63
Andrews: xxvii, 29, 30–31, 47, 50, 79–80, 112
Bill: 47
Mamie Lee: xxvii, 29, 30–31, 47, 50, 80, 112
Mamie Lee and Bill: 79
ANNEX 2, Requirements of Army Air Forces: 74
Army Reorganization Act of 1920: 44

Army: xii, xv, xvii, xx–xxi, 3, 6, 12, 21, 22, 24, 29, 36 43–46, 48, 63–64, 66, 74, 76–77, 124,
Plans Division: xiir
War Plans Division (WPD): 65–66
Arnold, Henry H. "Hap": 52, 64–65, 67–68, 71–73, 75, 96, 102, 121
Arrgentia Bay: 68
AT-5: 31
Atlanta Constitution: 45
Australia: xii, 18, 87, 89–91, 93–94, 103, 107, 109, 115
AWPD: 66–67, 70, 73, 77
AWPD-1: xii, 73–75, 77, 80
AWPD-2: 76
Axis: xxiii, 66, 68, 76

B-12A: 52
B-17: 37–38, 50, 52, 58, 65, 88, 90, 93–94, 97, 100, 108–109, 113–116, 129
B-18: 53
B-24: 103, 108–109 113–115, 125
B-25: 90, 97
B-26: 98, 100, 104, 115
B-29: xxiii
Baker Board: xx, 45
Barron Field: 6
Barton, Charles: 127–128
Battle: xxv, 54, 117, 129
of Britain: 54
of Saratoga: xxv
of the Bismarck Sea : 117
of the Bismarck Sea: 129
Benn, William G.: 98, 100–101
Bissell, Clayton L.: 66, 67, 73
Blanche Bay: 109
Bleasdale, Jack W.: 113, 125–126, 128–129
bombardier: 23, 35–36
bombardment: 22, 24–29, 32–34, 36–38, 43, 53, 57–58, 63–64, 67, 69–70, 72, 75, 77, 89, 108
command of 5th American Air Force: 117
commander: 111
expert: 54
force: 43, 75
group: 57, 77, 89

mission: 58
of operating bases: 69
operations: 72, 108
planes: 70
text: 70
Bomber Command: 103
bombing results: 96, 108
bombsights: 22, 23
Brereton, Lewis: xxiv
Brett, George: xxiv, 88
Brisbane (Kenney's headquarters): 91, 96, 98
Brooks Field: 6
Bullitt, William C.: 64
Buna: 92, 97, 103, 107, 116

CAMCO: 63
Camp Kiethly: 15
Camp Nichols: 10, 13, 17
Catalina: 115
Capp, William C.: 53
Central Aircraft Manufacturing Company: 63
Central Identification Laboratory (CIL): 128
Chanute Field: 52
Chennault, Claire L.: 29, 34–35, 37
Chicago Tribune: 77
China: 18
Churchill: 68
CIL: 128
Clark Field: 14
Coast Artillery Journal: 26
Command and General Staff School (CGSS): 38, 43–45, 50
Congress: xxi, 44–45, 64, 71
Conway: 49
Coral Sea: 87
Corregidor: 14
Currie, Lauchlin: 63

Daniel, Benton H.: 113, 126, 128–129
Davidson, Howard C.: 35
Department of Air Tactics and Strategy: 38
Department of National Defense: 46
DH-4: 8, 14, 21

Distinguished Service Crosses: 116
dive bombing: 93
doctrine: 43
Dollenburg, Fred P.: 95, 96
Doolittle, Jimmy: xx
Douhet, Giulio: xxii

Edgewood Arsenal: 25
Eglin: 100
Eighth Area Army Headquarters: 110
enemy air strength: 73
escort fighter: 69, 71, 73–74, 76
experimental jets: xviii

Fairchild, Florence: 29
  Muir S.: 29, 33, 63, 79
Father Poncelet: 126
Federal Aviation Commission: 44
Fenton Field: 89
Fifth Air Force: xii, 107–108, 129
Fifth Bomber Command: xii–xiii, 91, 95, 97–98, 107–108, 116–117, 125, 129
fighter planes: 25, 34, 35, 37, 74, 93, 108–109, 114–115, 129
First World War: xvi, xxiii
Fletcher, Allen S.: 15
FLYING FORTRESS: 53, 72
Fort: 6, 10, 12, 36–38
  Knox Maneuvers: 37
  Leavenworth: 38
  MacArthur: 36
  McKinley: 12
  Monroe: 30
  Sill: 6, 10
Foulois, Benjamin: xvii
Fourth Air Force: 91
Frank, Walter H.: 25, 65
  W. S.: 33

Gasmata: 108, 116
General Headquarters (GHQ): 27
General Headquarters Air Force: 45, 50
George, Harold L.: xii, 32–33, 38, 65, 127

GHQ Air Force: xxiv, 52–53, 64, 72
Gosport: 6
grand strategy: xxiii
Great War: xxi
Group Captain Garing of the Royal Australian Air Force (RAAF): 91
Guadalcanal: xiii, 93–94, 97–98, 107

Halsey, William F.: 107
Hamilton Field: 50, 52
Hansell, Haywood S.: xii, 30–31, 33, 35, 47, 57, 67–68, 70, 73–75
Hasher, David: 57
Hassemer, David: 113
Headquarters, Army Air Forces: 121
Higgins, Walter: 115, 125
high-altitude: xii, 38, 97
  daylight precision attacks: xii
  daylight precision bombing: 38
  formation bombing: 97
Hiller, Margaret: 9
Hitler: 50, 54, 64, 77
Hodge, C. L.: 78
  C. L. and Mary: 82, 127
  Mary: 78–82, 112–113, 127–128
Hodges, Courtney H.: 24
Holguin, Jose L.: 125, 128
Holloway, Bruce K.: 55–57, 59–60
Hoover, Herbert: xvii
Hopkins, Harry: 76
Horn Island: 89, 91
Howell, Clark: 45
Howell Commission: 45
Hoy, Bruce: 128–129
Huff-Daland Pegasus: 22
Huon Gulf: 107

Imamura: 110
Iron Range: 89, 103, 108, 113
Iwo Jima: xxiii

Jackson Field: 113
Japan: xv, xix, 63, 65–66, 82, 124, 127, 129

Japanese: xii, 58, 87, 89, 92–94, 98, 102, 107–111, 114–115, 124–126, 128–129
advance: xiii, 87, 126
armed forces: 87
attack: xxi
base at Rabaul: 88
convoy: 97, 116
Eighth Area Army Headquarters: 110
fighters: 93, 108, 114–115, 129
interrogations of prisoners: 128
operations: 107
planes: 90, 94
planes or ships: 58
positions: 88, 94
POW camp: 117, 127, 129
radar: 109
radio broadcast: 122
records: 124
shipping: 107
ships: 100
troops: 92
Java: 88, 90
Joint Board: 66

Kavieng: 129
Keifuku Maru: 115
Kelly Field: 48
Ken's Men: 103
Kenney, George C.: xii–xiii, xxiv, 24, 33–34, 90–91, 93–98, 100–103, 107–111, 113–117, 123
King, Ernest: xxiv
Kilbourne, C. E.: 45
Knerr, Hugh J.: 22–23, 48
Knox, Frank: 65
Kokoda: 92, 93, 95
Kuter, Ethel: 30–31, 79
Laurence S. "Larry": xii, 30, 33, 51–52, 67, 69, 71–75

Lae: 94, 107–108, 116
Langley Field: 22–23, 29–30, 32, 50–51, 62
large bombers: 72
Legion of Merit: 80, 121

Lend Lease Act: 64
Lend Lease aid to Britain, Russia, and China: 76
Liberators: 114
Liberty Bell trophy: 22
Liberty engines: 21
light bomber : 52
Lindberg, Allen: 113, 122
Lindbergh, Charles A.: 21
long-range: 37, 66, 72
  bomber: 37, 66
  fighter: 72
Lovett, Robert A.: 64, 75
low-altitude: 93, 97, 100, 103, 117
  attack: 103, 117
  skip bombing: 100
  tactics: 93
low-level attacks: 97

MacArthur, Douglas: xxiv, 87, 90, 94–96, 98, 107, 111, 115–117, 121
March Field: 22, 36, 52
Marines: xxiii
Marshall, George C.: 64
Marshall Islands: 87
Martin: 14, 22–23
  B-10 bomber: 23
  B-2 bombers: 22
  bombers: 14
Mather Field: 4
Maxwell Field: xviii, xxii, xxv, 32, 47
McAuley, Lex: 129
McCloy, John J.: 75
McCullar, Ken: 101, 103
McMurria, James A.: 115, 125
McNarney, Joseph: 33
Medal of Honor: xxvii, 115, 121, 127
Midway: 87
Milne Bay: 92–93, 115
Missing Personnel Investigation Unit (MPIU): 124
Mitchell, William "Billy": xx–xxii, 44, 21, 23–25
Moffett, William: xxi
Monihan, Gene: 127–129

Moresby: 89, 92, 94–95, 102–103, 107–108, 113, 115, 126
Morrow Board: xx
MPIU: 124–125

Nagaaki Kawai: 110
Nagumo: xxiv
National Air Races: 22
National Defense Act of 1920: xxi
National Museum and Art Gallery: 128
national security: 43
Navy: xviii, xx–xxi, xxiv, 21, 43, 46, 76–77, 93
NBS-l: 14
Netherlands East Indies: 87, 107
New Britain: 87, 108, 125–126
New Guinea: xiii, 87–88, 93–96, 103, 107, 110, 117
night bomber: 14
night raids: 72
Norden bombsight: 23, 38
Northeast Area Command: 91
Norton, Patrick: 98, 100, 103–104, 116–117

Olds, Robert: 24–26, 45–46, 50, 63
Order of Our Lady of the Sacred Heart : 126
Oval Office: 121
Overturf, Samuel: 1
Owen Stanley Mountains: 88, 92–93, 111, 114

P-36: 55
P-38: 93
P-51: xxiii
Pacific: xxi, 58, 87, 95, 107, 128
Packard engine: 22
Papua, New Guinea: 97, 128
Papuan crisis: 94
Papuan Peninsula: 87, 89, 92–93, 102
Peabody, Hume: 32, 34
Pearl Harbor: xxi, xxiii–xxiv, 58, 77, 79
Pearson, Bob: xvii, 1, 7, 45, 80, 112
Pentagon: 70
Perakos, John: 116
Pettit Barracks: 15
Philippine: xvi, xxiv, 10, 12, 19, 47, 76, 87–88, 90, 126–12

Command: 127
Islands: 10, 12
tour: 19
Ryukyus Command: 126
Plans Division: 63, 121
Port Moresby: 89, 91, 93
Post Field: 6, 8, 18
Potter, Hosea: 9
POW camp: 126
pre-World War II Air Corps: 36
pre-World War II aviation officers: 37
pursuit: xi, 22–27, 29, 32–35, 37–38, 43, 55, 72, 74, 76, 90

Quinones, Alphonse D.: 125–126

RAAF: 91, 103, 109
Catalinas: 109
combat units: 91
Rabaul: 87, 94, 97–98, 100, 107–110, 113–116, 124–126, 129
RAF: 72
Rainbow-5: 65, 67
Ramey, Howard: 117
Rapopo: 109
realities of warfare: 38
Regular Army Air Service: 29
Regular Army: 44
Reinburg, George E. A.: 12
reinforcements: 93
reorganization of the air forces: 80
Rickenbacker, Eddie: 95
Robbins, Charlie: 59
Robinson, Gerald: 53
Roosevelt, Franklin D.: 44, 64–65, 68, 121
Royal Air Force: 67, 100

Salamaua: 94, 107
Sananada: 108
Sanxter, Donald: 112, 114
Saville, Gordon: 33–34
Schneider: 69
School for Aerial Observers: 6
Scott: 80–81

Second World War: xxv, 65
Secretaries of War and Navy: 65
Senate: 81
Seven-Mile Field: 103
Shelton: 11, 17–18, 48–50
  Douglas: 18, 48
  Harvey and Virginia: 11, 49
Shelton's 31st Infantry: 12
Silver Star: 96
Simpson Harbour: 109
Sino-Japanese War: 72
skip bombing: 93, 100, 101, 117
Solomon Islands: 87, 93, 109–110
South China Sea: 87
South Pacific: xiii, 93, 107
Southeast Asia: xvi
Southwest Pacific: xi–xiii, xix, xxvii, 83, 87
Spaatz, Carl A. "Tooey": 35–36, 55, 63, 65, 67
SR&D: 125–126
St. George's Channel: 109
State Department: 78
Status Review and Determination Section: 124
Stimson, Henry L.: 64–65, 75, 77
Stratemeyer, George: 50, 121
surprise attack: 58

test model fighters: 73
The Air Force course: 29
The Great Depression: xix
Thomas: 14
Tobera: 109
torpedo bombers: 58
Torrens Creek: 89
Townsville: 88, 90–91, 93, 100–101
transport squadrons: 89
Travis, William: 95–96, 103, 116–117
Tripler Hospital: 53
Trobriand Islands: 115
two-ocean war: 65

US: 10, 32, 93–94, 100, 127
  32d Division: 94

Armament School: 100
Army: 10, 32
Marines: 93
Navy: 127

V Corps: 24
Van Atta, Lee: 125–126
Vandenberg, Hoyt S.: 33, 69, 74
Vunakanau: 109, 115

Wake: 87
Walker: xi–xiii, xv–xvi, xviii–xxi, xxiii–xxiv, 1, 4, 6–8, 12–19, 21–38, 43–59, 63–65, 67, 69–82, 87–91, 93–98, 100–104, 108–117, 121–123, 125–129
Douglas: 128
Emma: 1, 3
Ken: xxvii, 1, 21, 52–53, 82, 101, 103, 117
Kenneth: vii, xi, xiii, xv, xviii, xxi–xxii, 30, 50, 59, 63, 82–83, 116, 112, 121, 125
Marguerite Potter: 9–10, 17–18, 29–31, 49–50, 59, 81, 83, 112
Wallace: 1
War College: 54
War Department: 45–46, 52, 65, 67–68, 74, 76–77, 80, 122, 127
attitude: 45
General Staff: 65, 67
organization: 80
War Plans Division: 45, 68, 80
War Plans Group: 65
War Room: 71
Washington: xii, xv–xvi, 58, 63, 68, 78–80, 82–83, 90, 95, 102, 111
Washington Times-Herald: 77
Webster, Robert M.: 32–33
Wedemeyer, Albert C.: 66
Wewak, New Guinea: 125
Wheeler Field: 54, 58, 113
Whitacre, William: 114
Whitehead, Enis C.: xii, xxiv, 87, 91, 93–96, 108, 111, 117
Wide Bay: 115, 126, 129
Wilson, Donald: 26, 32–33, 38, 43, 45–46, 97, 109–111, 117
Wirt: 11
World War I: xvii–xix, xxii, 3, 21, 23, 27, 48
World War II: xiii, xviii, xx, 23, 33

WPD: 66–67, 74

Yamamoto: xxiv
Young, Sir Mark: 127–128

Zero: 88, 90, 114, 126